CAMBRIDGE
UNIVERSITY PRESS

Physics

for Cambridge International AS & A Level

PRACTICAL WORKBOOK

Graham Jones, Steve Field, Chris Hewlett & David Styles

CAMBRIDGE
UNIVERSITY PRESS

University Printing House, Cambridge CB2 8BS, United Kingdom

One Liberty Plaza, 20th Floor, New York, NY 10006, USA

477 Williamstown Road, Port Melbourne, VIC 3207, Australia

314–321, 3rd Floor, Plot 3, Splendor Forum, Jasola District Centre, New Delhi – 110025, India

103 Penang Road, #05-06/07, Visioncrest Commercial, Singapore 238467

Cambridge University Press is part of the University of Cambridge.

It furthers the University's mission by disseminating knowledge in the pursuit of education, learning and research at the highest international levels of excellence.

www.cambridge.org
Information on this title: www.cambridge.org/9781108793995

First published 2018
Second edition 2020

20 19 18 17 16 15 14 13 12 11 10 9 8 7 6

Printed in Poland by Opolgraf

A catalogue record for this publication is available from the British Library

ISBN 978-1-108-79399-5 Practical Workbook

Additional resources for this publication at www.cambridge.org/9781108793995

....

....

> Contents

⟩ Introduction

Practical work is very important in physics. Many discoveries in the physical world have only been made because practical work has enabled a theory to be proved beyond reasonable doubt, or has shown that current theories or ideas need to be changed. Even today, many of the principles you will learn as part of your course may only be approximations, and physicists realise that there are still many discoveries to be made. Your generation will probably provide insights that will enhance our understanding of the physical world and improve our current theories. But do remember that the work of experimental and theoretical physicists can only be proven to be correct with suitable experimental work. Such experimental work may be on an astronomical scale, for example, to establish what exactly gravity is, or on a microscopic scale, for example, to establish how particles like electrons or atoms can be considered to have wave properties.

It is generally acknowledged that good quality practical work develops a range of skills, knowledge and conceptual understanding. Such skills, involving genuine enquiry, are valuable to the science community as a whole, as well as to physics, and are useful in other areas such as industry and business. By learning how to approach a practical problem, how to plan an investigation, how to take adequate measurements and how to analyse your results, you will be developing skills that you are very likely to make good use of in later life.

You may, initially, be worried because you have done little practical work before you started this course, or perhaps practical work has simply been following instructions to collect specified data using unfamiliar equipment, following stated procedures that perhaps you did not understand. This course is designed to help you improve your practical skills, and also to help you prepare for your practical examinations. The skills you will need for these are developed during this course as you progress through the workbook. You will be planning investigations for yourself, taking measurements and analysing your own results. You should take ownership of these results and use your practical time well.

An investigation does not always go well. However, some important advances, such as superconductivity, were made when physicists did not just give up when an experiment did not work; instead they analysed the unexpected results and then they thought carefully about problems with the apparatus. You can do the same, so that you can learn from investigations that do not work as well as from those that do. This requires thought, but hopefully it will stimulate your interest and determination, as well as helping you to develop valuable skills.

You will find guidance on some of the practical skills in Chapters P1 and P2 of the Cambridge International AS & A Level Physics Coursebook in this series. You may like to use these chapters as an introduction, or as a reference.

Above all, enjoy your learning and practical work; you may be surprised how enjoyable it can be!

> Safety

Working safely in a physics laboratory is an essential aspect of learning which characterises practical work.

Always listen carefully to instructions and carefully follow written instructions and codes of conduct.

If unsure about any aspect of your practical work, ask your teacher. If designing your own investigation, ask your teacher to check your plan before carrying it out.

Many safety issues in a physics laboratory concern the prevention of damage to the equipment.

Working with water	Place all the apparatus in a tray so that any spillage does not affect paperwork. If working with hot or boiling water, use tongs to handle containers such as beakers.
Using a liquid-in-glass thermometer	Place the thermometer securely on the bench when not in use, so that it does not roll off the bench. If a thermometer breaks, inform your teacher immediately. Do not touch either the broken glass or the liquid from inside the thermometer.
Loading thin materials such as wires	Wear safety goggles in case of fracture of the wire. Beware of falling weights if the wire fractures and place a cushion or similar object on the floor.
Connecting electrical components	Do not exceed the recommended voltage for the component: for example, a 6 V lamp.
Toppling retort stands	If a stand is moving or in danger of toppling, secure it to the bench using a G-clamp.
Rolling objects such as cylinders	Place a suitable object such as a box to collect the object so that it does not fall to the floor or affect somebody else's experiment.
Dry cells such as 1.5 V batteries	Do not connect the terminals of the cell to each other with a wire.
Using sharp blades or pins	Tape over sharp edges; keep points of pins downwards, away from eyes.

Table S.1: Safety issues in a physics laboratory.

› Practical skills

Collection of data

There are several stages needed to collect experimental data. You will need to take accurate measurements from a variety of different apparatus.

You will need to:

- follow instructions both written and in a diagram form

- use practical apparatus correctly

- use practical apparatus safely

- use both analogue and digital measuring instruments

- consider methods to increase the accuracy of measurements, such as timing over a multiple number of oscillations, using a fiducial marker, or using a set square or plumb line

- construct circuits from circuit diagrams ensuring that ammeters and voltmeters are connected correctly, understanding the importance of the polarity of a power supply

- use a signal generator

- use a cathode ray oscilloscope to include the determination of the potential difference from the y-axis and the time from the x-axis.

In practical activities where a straight-line relationship is expected, the *minimum* number of measurements to be taken should be six. More readings should be taken for a curved trend.

You will need to decide on the range over which you will take readings. The measurements should cover as large a range as possible, with sensible intervals between the readings in the range.

You should also consider repeating readings and determining the mean value.

Presentation of results

Measurements and observations made during an experiment need to be recorded in ways that are easy to follow.

Initially, results should be recorded in a table. The table of results should be planned before the experiment is performed.

The columns of the results table should include space for measurements that are taken and space for values which are going to be calculated from the measurements. Each column heading should include both the quantity and the unit. There should be a distinguishing mark (usually a forward slash '/') to separate the quantity and the unit, for example, length / m, length (cm), L / mm, etc.

Measurements taken in an experiment should be recorded to the same precision as the measuring instrument. For example, when using a ruler to measure length, measurements are made to the nearest millimetre, so a length of ten centimetres should be recorded as 10.0 cm. In timing experiments using a stopwatch, times can either be recorded to the nearest 0.01 s or rounded to the nearest 0.1 s.

Where calculated values are recorded in your results table, remember to label the column heading with both the quantity and the unit. For example, in an experiment to measure time, t, you may need to calculate time squared, t^2. In this case, the column should be labelled t^2 / s^2.

The calculated values should be recorded to the same number of significant figures as the raw data used to determine the calculated value.

For example, consider the diameter d of a wire. The calculated column may require d^2 to be determined.

d / mm	d^2 / mm^2	d^2 / mm^2 to two significant figures
0.27	0.0729	0.073
0.28	0.0784	0.078
0.29	0.0841	0.084

Table P.1: Example results table showing calculated values.

As the values in Table P.1 indicate, a change in the second significant figure affects the second significant figure in the calculated value.

Example table of results

In an experiment, the potential difference V across a wire and the current I flowing through the wire are measured for different lengths L of the wire. The resistance of the wire is then calculated as shown in Table P.2.

Each column heading has a quantity and a unit.

All the length values have been measured to the nearest millimetre.

L / cm	V / V	I / mA	R / Ω
22.2	11.6	73.2	158
30.0	11.6	60.2	193
40.0	12.0	52.8	227
49.7	12.0	45.2	265
59.3	12.2	40.4	302
68.8	12.4	37.6	330

Since both V and I have been recorded to three significant figures, R is also recorded to three significant figures.

All the V values have been measured to the nearest 0.1 V and all the I values have been recorded to the nearest 0.1 mA.

Table P.2: Example results table.

At A Level, you will need to understand how to calculate logarithms. For example, consider calculating the logarithm of L for the value of 22.2 cm from Table P.2.

L / cm	log (L / cm)	log (L / cm) to three decimal places
22.1	1.344 39	1.344
22.2	1.346 35	1.346
22.3	1.348 30	1.348

Table P.3: Example results table showing calculated logarithmic values for L.

In Table P.3, the number of significant figures (three) in the raw data (L) corresponds to the number of decimal places in the calculated logarithmic value. In general, logarithmic values should be given to the same number of decimal places as the *least* number of significant figures in the measured quantities.

Note the number before the decimal place in a logarithmic quantity is a place value. For example, in Table P.3, 22.2 cm can be written as 222 mm. In this case log (L / mm) = 1.346: only the number *before* the decimal place has changed.

Graphs

It is useful to follow a set procedure every time you plot a graph. It is also very useful to draw graphs in pencil so that it is easy to make changes.

Label the axes

The independent variable should be plotted on the x-axis and the dependent variable should be plotted on the y-axis. Each axis should be labelled in the same way as the column heading in the table, using a quantity and a unit.

Add a scale to each axis

The points should occupy more than half the graph grid in each direction. Scales should be simple so that points can easily be plotted and information can easily be read from the graph. Use simple proportions for each 1 cm or 2 cm square, such as 1, 2, 4, 5, 10, 20 or 50. The scales should be labelled at least every two large squares.

Plot the points

All the data values from the table should be plotted on the graph grid (not in the margin area by adding lines to the graph). The plots should be clear and not too thick. It is advisable to check any plots that do not follow a trend.

Draw the best-fit line

Use a transparent 30 cm ruler so that you can easily see the points you have plotted and ensure any line you have drawn is not going to have a join where two rulers have been used. When drawing the best-fit line (or curve) there should be a reasonable balance of points about the line. The line should not be too thick.

At A Level you will also need to add error bars and draw worst acceptable lines.

Figure P.1 has been plotted from the data in Table P.2.

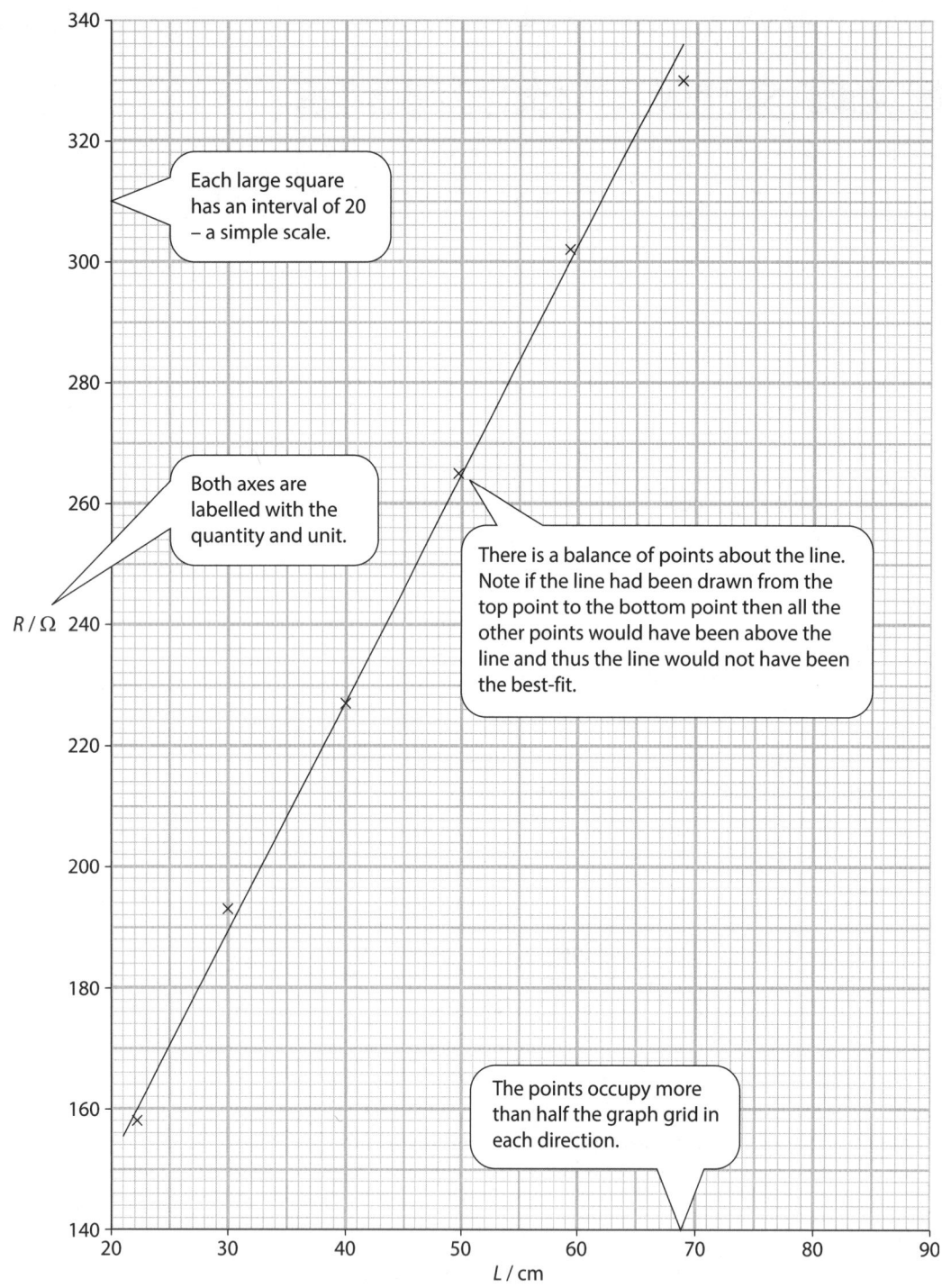

Figure P.1: Example graph with line of best fit.

Sometimes you may have a curved trend and you will need to be able to draw a tangent to the curve at a point on the curved trend.

Interpretation of graphs and conclusions

It is expected that you can determine the gradient and y-intercept from a graph. All your working should be shown.

To find the gradient, choose two points (x_1, y_1) and (x_2, y_2) that lie on the best-fit line. The two points should be at least half the length of the line apart. It is helpful to try to use points where the best-fit line crosses grid lines. You should only use plots from the table of results if the plots are clearly on the best-fit line. In Figure P.1 the plot from Table P.2 (22.2, 158) does *not* lie on the best-fit line.

$$\text{gradient} = \frac{\Delta y}{\Delta x} = \frac{y_2 - y_1}{x_2 - x_1}$$

Using this method, negative gradients are identified.

There are two methods to determine the y-intercept:

a   reading the value directly from the y-axis when $x = 0$

b   using the calculated gradient and substituting a point from the best-fit line into the equation of a straight line.

The simplest way to do method **b** is to use one of the points from the gradient calculation:

$$y\text{-intercept} = y_2 - \text{gradient} \times x_2 \text{ or } y\text{-intercept} = y_1 - \text{gradient} \times x_1$$

In Figure S1 the x-axis starts at 20 so the substitution method must be used.

The gradient and y-intercept values that you have determined can then be used to determine other values from a given relationship. For Figure P.1 and Table P.2, the given relationship between R and L may be:

$$R = \frac{4\rho L}{\pi d^2}$$

where ρ is the resistivity of the material and d is the diameter of the wire. You may need to measure the diameter and then determine ρ. It is useful to identify the gradient from the equation:

$$\text{gradient} = \frac{4\rho}{\pi d^2}$$

and then to rearrange the equation with ρ as the subject:

$$\rho = \frac{\text{gradient} \times \pi d^2}{4}$$

Take care with powers of 10. In Tables P.2 and P.3, $L\,/\,\text{cm}$ could have been written as $L\,/\,10^{-2}\,\text{m}$. The final answer should be given to an appropriate number of significant figures with a correct unit.

Identifying limitations and suggesting improvements

Experimental procedures are not perfect. You should be able to identify limitations in the experimental design and suggest ways in which the procedure might be improved.

Planning

You should develop skills to design appropriate practical activities.

You will need to:

- identify the independent variable (the quantity that will be changed) and the dependent variable (the quantity that will be measured)

- consider how you can make the experiment or investigation a fair test by controlling variables, i.e. identifying the quantities to be kept constant

- select appropriate apparatus

- select appropriate measuring instruments

- describe, in detail, the techniques needed for the proposed experiment, including drawing a labelled diagram

- describe any safety precautions explicitly needed for the experiment

- explain how a relationship may be tested; for example, suggest an appropriate graph to plot by identifying quantities to plot on each axis

- explain how the data collected may be analysed.

When testing relationships, remember that a directly proportional relationship will produce a straight line passing through the origin. For a linear relationship that is not directly proportional, you would expect a straight line *not* passing through the origin.

At A Level you are expected to be able to analyse relationships using both logarithms to the base 10 (lg) and natural logarithms (ln).

For example, if the relationship between P and Q is:

$$Q = kP^n$$

where k and n are constants, then the linear form of this equation may be found by taking logs of both sides:

$$\lg Q = n \lg P + \lg k$$

For this relationship to be valid, a graph of $\lg Q$ (y-axis) against $\lg P$ (x-axis) should be plotted.

$$n = \text{gradient and since } y\text{-intercept} = \lg k, \text{ then } k = 10^{y\text{-intercept}}$$

Similarly, if the relationship between S and T is:

$$T = ke^{-wS}$$

where k and w are constants, then the linear form of this equation may be found by taking natural logs of both sides:

$$\ln T = -wS + \ln k$$

For this relationship to be valid, a graph of $\ln T$ (y-axis) against S (x-axis) should be plotted.

$$w = -\text{gradient and since } y\text{-intercept} = \ln k, \text{ then } k = e^{y\text{-intercept}}$$

Treatment of uncertainties

Measurements should include an estimate of the absolute uncertainty. This is an estimate of the maximum difference between the actual reading and the true value. For example, in Table P.2 V may have been measured to the nearest 0.2 V. The first reading of V could be written as (11.6 ± 0.2) V.

If readings are repeated, then the absolute uncertainty is half the range of the values.

Often it is useful to write uncertainties as fractional uncertainties or percentage uncertainties:

$$\text{fractional uncertainty} = \frac{\text{absolute uncertainty}}{\text{(average) value}}$$

$$\text{percentage uncertainty} = \frac{\text{absolute uncertainty}}{\text{(average) value}} \times 100\%$$

It is often necessary to combine uncertainties. There are some simple rules:

- When adding or subtracting quantities, add the absolute uncertainties.

- When multiplying or dividing quantities, add the percentage uncertainties.

- There is a special case for power terms. For example, if $Q = P^n$

percentage uncertainty in $Q = n \times$ percentage uncertainty in P.

In a table of results, it may be necessary to indicate the absolute uncertainty in a calculated quantity. For example, in the first row of Table P.2, it could be that V was (11.6 ± 0.2) V and I was (73.2 ± 0.2) A. One way of determining the absolute uncertainty in R is to determine the percentage uncertainty in R first.

$$\text{percentage uncertainty in } R = \left(\frac{0.2}{11.6} + \frac{0.2}{73.2} \right) \times 100 = 2\%$$

Then use the percentage uncertainty to determine the absolute uncertainty in R.

$$\text{absolute uncertainty in } R = \frac{2}{100} \times 158 = 3$$

There are alternative methods using maximum and minimum values. Using the data:

$$\max R = \frac{(11.6 + 0.2) \text{ V}}{(73.2 - 0.2) \text{ mA}} = 161.6 \ \Omega \text{ and}$$

$$\min R = \frac{(11.6 - 0.2) \text{ V}}{(73.2 + 0.2) \text{ mA}} = 155.3 \ \Omega$$

Each of these values gives an absolute uncertainty of about 3 and half the range is also about 3.

You can show the uncertainties of the measurements on a graph by drawing error bars. The length of each error bar either side of the plot corresponds to the absolute uncertainty in the quantity.

Having drawn the error bars, a worst acceptable line may be drawn. This is the steepest or shallowest line that, when drawn, passes through all the error bars.

The methods used to determine the gradient and y-intercept of the best-fit line can also be used to find the gradient and y-intercept of the worst acceptable line. Then you can find the absolute uncertainty in these values.

absolute uncertainty in gradient = gradient of best-fit line − gradient of worst acceptable line

absolute uncertainty in y-intercept = y-intercept of best-fit line − y-intercept of worst acceptable line

Using the data from the earlier example and assuming the uncertainty in V is ± 0.2 V and the uncertainty in I is ± 0.2 mA, the results become those shown in Table P.4.

L / cm	V / V	I / mA	R / Ω
22.2	11.6 ± 0.2	73.2 ± 0.2	158 ± 3
30.0	11.6 ± 0.2	60.2 ± 0.2	193 ± 4
40.0	12.0 ± 0.2	52.8 ± 0.2	227 ± 5
49.7	12.0 ± 0.2	45.2 ± 0.2	265 ± 6
59.3	12.2 ± 0.2	40.4 ± 0.2	302 ± 6
68.8	12.4 ± 0.2	37.6 ± 0.2	330 ± 7

Table P.4: Example data taken from Figure P.2.

Plot the graph with error bars for each value of resistance. Draw the best-fit line and then the worst acceptable line. Figure P.2 shows the error bars and the (shallowest) worst acceptable line.

The lines should be labelled and it is usual to show the worst acceptable line as a dashed line. The worst acceptable line should pass through all the error bars unless there is an anomalous plot. The actual worst acceptable line should clearly be seen to pass through each error bar.

The gradient, including the absolute uncertainty, may then be determined:

$$\text{gradient of best-fit line} = \frac{322 - 208}{65 - 35} = \frac{114}{30} = 3.80$$

$$\text{gradient of worst acceptable line} = \frac{326 - 178}{68 - 27} = \frac{148}{41} = 3.61$$

> **TIP**
>
> Use data points from the worst acceptable line.

Uncertainty in gradient = 3.8 − 3.61 = 0.19

Gradient = 3.80 ± 0.19 (Ω cm^{-1}). The units are useful at this stage for any further calculation.

Since there is a false origin, the y-intercept, including the absolute uncertainty, may be determined from the gradient calculations:

y-intercept of best-fit line = 322 − 3.80 × 65 = 75

y-intercept of worst acceptable line = 326 − 3.61 × 68 = 81

Uncertainty in y-intercept = 81 − 75 = 6

y-intercept = (75 ± 6) Ω

> **TIP**
>
> Use the gradient from the worst acceptable line and a data point from the worst acceptable line.

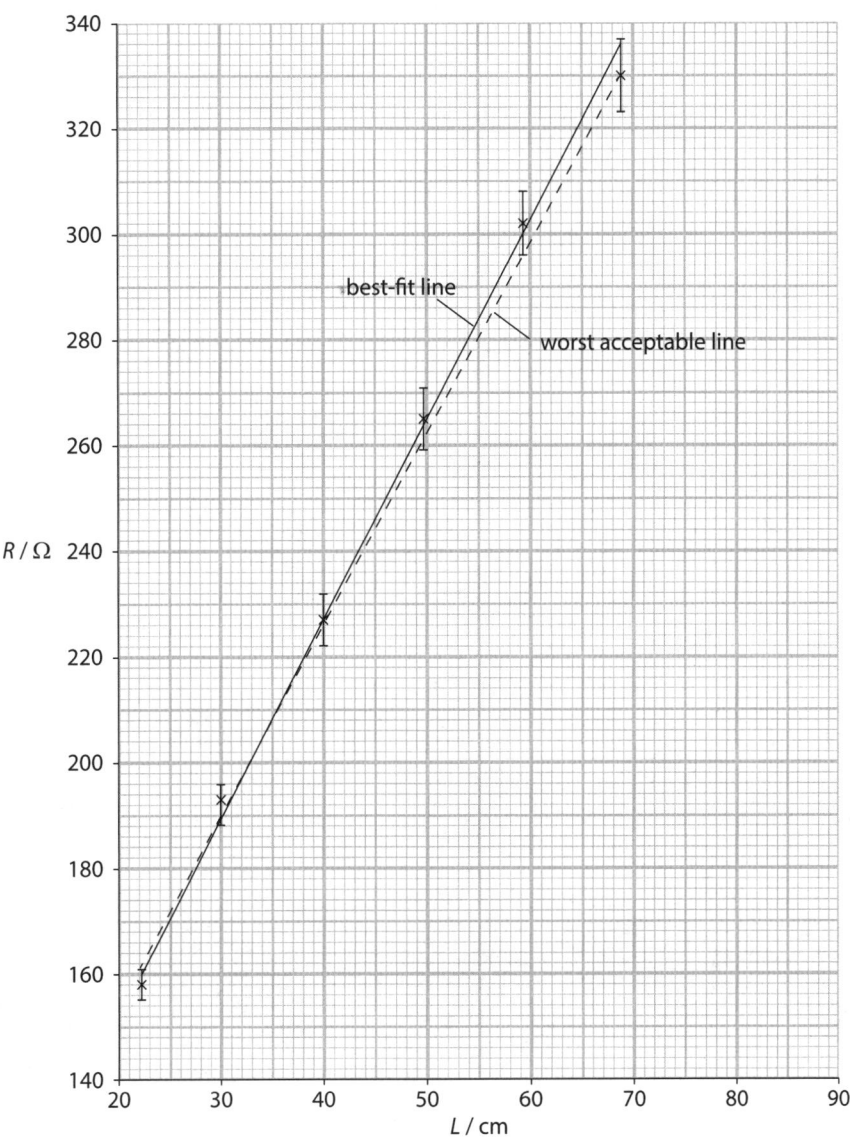

Figure P.2: Example graph with line of best fit and worst acceptable line.

> How to use this series

This suite of resources supports students and teachers following the Cambridge International AS & A Level Physics syllabus (9702). All of the books in the series work together to help students develop the necessary knowledge and scientific skills required for this subject. With clear language and style, they are designed for international learners.

The coursebook provides comprehensive support for the full Cambridge International AS & A Level Physics syllabus (9702). It clearly explains facts, concepts and practical techniques, and uses real-world examples of scientific principles. Two chapters provide full guidance to help students develop investigative skills. Questions within each chapter help them to develop their understanding, while exam-style questions provide essential practice.

The workbook contains over 100 exercises and exam-style questions, carefully constructed to help learners develop the skills that they need as they progress through their Physics course. The exercises also help students develop understanding of the meaning of various command words used in questions, and provide practice in responding appropriately to these.

This write-in book provides students with a wealth of hands-on practical work, giving them full guidance and support that will help them to develop all of the essential investigative skills. These skills include planning investigations, selecting and handling apparatus, creating hypotheses, recording and displaying results, and analysing and evaluating data.

The teacher's resource supports and enhances the questions and practical activities in the coursebook. This resource includes detailed lesson ideas, as well as answers and exemplar data for all questions and activities in the coursebook and workbook. The practical teacher's guide, included with this resource, provides support for the practical activities and experiments in the practical workbook.

Teaching notes for each topic area include a suggested teaching plan, ideas for active learning and formative assessment, links to resources, ideas for lesson starters and plenaries, differentiation, lists of common misconceptions and suggestions for homework activities. Answers are included for every question and exercise in the coursebook, workbook and practical workbook. Detailed support is provided for preparing and carrying out for all the investigations in the practical workbook, including tips for getting things to work well, and a set of sample results that can be used if students cannot do the experiment, or fail to collect results.

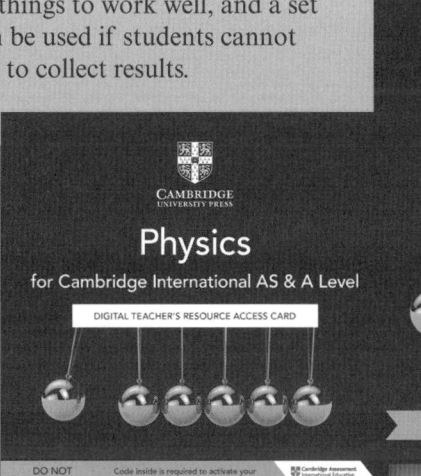

> How to use this book

Throughout this book, you will notice lots of different features that will help your learning. These are explained here.

CHAPTER OUTLINE

These appear at the start of every chapter to introduce the learning aims and help you navigate the content.

TIP

The information in these boxes will help you complete the exercises, and give you support in areas that you might find difficult.

Exercises

Appearing throughout this book, these help you to practise skills that are important for studying AS & A Level Physics.

KEY EQUATION

Equations which will help you to complete the exercises are provided throughout the book.

KEY WORDS

Key vocabulary is highlighted in the text when it is first introduced. Definitions are then given in the margin, which explain the meanings of these words and phrases.

You will also find definitions of these words in the Glossary at the back of this book.

Using apparatus

CHAPTER OUTLINE

This chapter relates to Chapter 1: Kinematics: describing motion, Chapter 7: Matter and materials and Chapter 8: Electric current, potential difference and resistance, in the coursebook.

In this chapter, you will complete investigations on:

- 1.1 Determining the density of water

- 1.2 Determining the spring constant of a spring

- 1.3 Determining the resistance of a metal wire

- 1.4 Determining the average speed of a cylinder rolling down a ramp.

Practical investigation 1.1: Determining the density of water

Density is defined as mass ÷ volume or, expressed in symbols:

$$\rho = \frac{m}{V}$$

The standard unit for density in the SI system of units is kg m^{-3}. 1000 kg m^{-3} = 1 g cm^{-3}.

KEY EQUATION

density $\rho = \dfrac{m}{V}$

YOU WILL NEED

Equipment:
- metre rule • 30 cm ruler • 250 cm^3 beaker • Vernier or digital callipers.

Access to:
- jug of water • top-pan balance.

Safety considerations

- Make sure you have read the Safety advice at the beginning of this book and listen to any advice from your teacher before carrying out this investigation.

- Clear any spillages of water.

Part 1: Determining density from single mass and volume measurements

Method

1   Place an empty 250 cm³ beaker on a top-pan balance. Record the reading on the balance.

Mass of empty beaker = ... g

2   Pour some water into the beaker until the water level is approximately 180 cm³.

Estimate the volume of the water.

Estimated volume of water V = ... cm³

3   Record the new reading on the balance.

Mass of beaker and water = ... g

Data analysis

a   Calculate m using:

m = mass of beaker and water – mass of beaker

m = ... g

b   Calculate the density ρ of water using your measurements.

ρ = ... g cm⁻³

Part 2: Using a graph to find density

Method

1   Place an empty 250 cm³ beaker on a balance. Record the reading on the balance in the Results section.

2   Pour some water into the beaker until the water level is approximately 50 cm³.

3   Record the new balance reading in Table 1.1 in the Results section.

4   The water in the beaker has a diameter d and height h.

   i   Measure d using the 30 cm ruler and record your measurement in the Results section.

   ii   Measure h using the metre rule and record your measurement in Table 1.1.

5   Change the amount of water in the beaker and take a series of readings of the mass of the beaker and the water and the height h. Record your results in Table 1.1.

Results

Mass of beaker = ... g d = ... cm

Mass of beaker and water / g	m / g	h / cm

Table 1.1: Results table.

Analysis, conclusion and evaluation

a   Calculate m for each of your readings using

m = mass of beaker and water − mass of beaker

Record your values for m in Table 1.1.

b   Plot a graph of m on the y-axis against h on the x-axis using the graph grid.

c   Draw the straight **line of best fit**.

d   Determine the **gradient** of this line.

KEY EQUATION

$$\mathbf{gradient} = \frac{\text{change in } y}{\text{change in } x} = \frac{\Delta y}{\Delta x}$$

Gradient = ...

e   Extension question. The volume v of a cylinder with diameter d and height h as shown in Figure 1.1 is given by:

$$V = \frac{\pi d^2 h}{4}$$

Using $\rho = \frac{m}{V}$ and $V = \frac{\pi d^2 h}{4}$, show that $m = \frac{\rho \pi d^2 h}{4}$

....

....

....

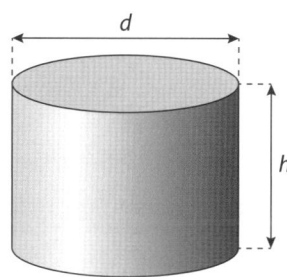

Figure 1.1: A cylinder.

f   Extension question. Show that the gradient of the graph will be $\frac{\rho \pi d^2}{4}$

....

....

....

TIP

π, ρ, d and 4 are all constant.

g   Determine ρ using:

$$\rho = \frac{4 \times \text{gradient}}{\pi d^2}$$

$\rho = $ .... g cm^{-3}

h   Suggest two advantages of using digital callipers instead of a ruler to measure d.

....

....

Practical investigation 1.2: Determining the spring constant of a spring

The **spring constant** is defined as force per unit extension, or expressed in symbols:

$$k = \frac{F}{e}$$

The stiffness of a spring is its resistance to deformation when a load is applied. The stiffer the spring, the greater the value of k.

The standard unit for spring constant in the SI system of units is N m^{-1}. $100 \text{ N m}^{-1} = 1 \text{ N cm}^{-1}$.

KEY EQUATION

spring constant $k = \dfrac{F}{e}$

YOU WILL NEED

Equipment:

- expendable steel spring • 100 g mass hanger • 0–10 N newton-meter
- 30 cm ruler • four 100 g slotted masses • two stands • two bosses
- two clamps • G-clamp.

Safety considerations

- Make sure you have read the Safety advice at the beginning of this book and listen to any advice from your teacher before carrying out this investigation.

- If the stand moves or tilts it may be necessary to secure it to the bench using the G-clamp.

Part 1: Determining the spring constant from the measurement of an extension and the calculation of a load

Method

1 Measure the length x_0 of the coiled section of an unextended spring as shown in Figure 1.2 and write your answer in the Results section.

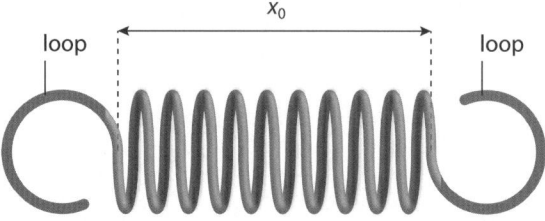

Figure 1.2: Spring with loops at ends.

2 Suspend the spring from the rod of a clamp.

Attach 500 g from the bottom of the spring as shown in Figure 1.3.

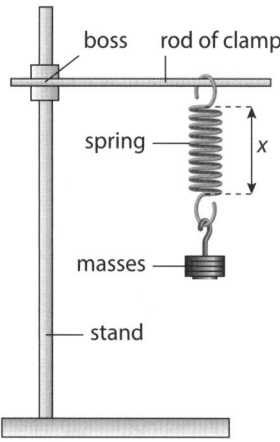

Figure 1.3: Spring suspended from rod with weights.

3 Measure the length of the coiled section x of the extended spring as shown in Figure 1.3 and write your answer in the Results section.

4 Record the masses of the mass hanger and each of the slotted masses separately to the nearest 0.1 g in Table 1.2 in the Results section.

Results

$x_0 = $ .... cm $x = $ .... cm

Mass / g				
mass hanger	mass 1	mass 2	mass 3	mass 4

Table 1.2: Results table.

Analysis, conclusion and evaluation

a   Calculate the extension e of the spring using:

$e = x - x_0$

Give your answer in metres.

$e =$ .... m

b   Calculate the total value m of the mass hanger and the 100 g slotted masses.
Give your answer in kg.

$m =$ .... kg

c $F = m \times g$, where g is acceleration of free fall equal to 9.81 m s^{-2}.

Calculate the spring constant k using $k = \dfrac{F}{e}$

$k =$ .... N m^{-1}

d   The following will contribute to the **uncertainty** in x:

- both ends of the rule must be viewed at the same time
- the exact positions where the coiled section starts and ends may be unclear.

List *two* further sources of uncertainty.

....

....

e   Calculate the mean value of a 100 g slotted mass using the values in Table 1.2.

Mean value = ... g

f   Calculate the uncertainty in the value of a 100 g mass from the half **range**
given by:

$$\frac{\text{largest value of mass} - \text{smallest value of mass}}{2}$$

Uncertainty = ... g

KEY WORDS

uncertainty (also absolute uncertainty): an estimate of the spread of values around a measured quantity within which the true value will be found

---

KEY EQUATION

$\text{uncertainty} = \dfrac{1}{2}$ (maximum reading − minimum reading)

KEY WORD

range: the difference between the largest value and the smallest value of a measurement

Part 2: Using a newton-meter to measure force

Safety considerations

- Make sure you have read the Safety advice at the beginning of this book and listen to any advice from your teacher before carrying out this investigation.

- Take care when moving the bottom clamp because the spring balance and/or the spring could slide off the end of the rod.

Method

1 Set up the apparatus as shown in Figure 1.4. Use the same spring as you used in Part 1.

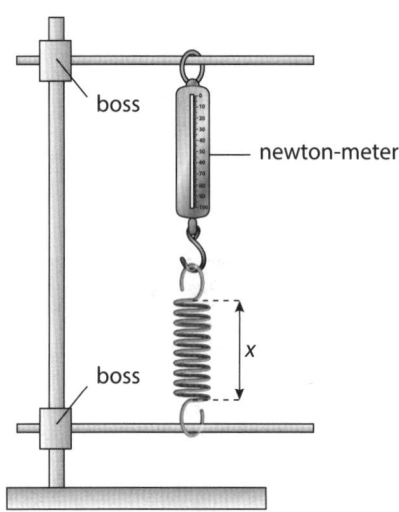

Figure 1.4: Spring between two rods, with newton-meter.

2 Move the bottom clamp vertically to different positions. Take a series of readings of F and x.

Record your data in Table 1.3 in the Results section.

> **TIP**
>
> The newton-meter will record a reading of force F in newtons.

Results

F / N	x / cm	e / cm

Table 1.3: Results table.

Analysis, conclusion and evaluation

a Calculate the extension e of the spring and add these values to Table 1.3.

b Plot a graph of e on the y-axis against F on the x-axis using the graph grid on the next page.

c Draw the straight line of best fit.

d Determine the gradient of this line.

Gradient = ...

e Extension question. Show that:

$$k = \frac{1}{\text{gradient}}$$

....

....

f Extension question. Determine k from your gradient.

$k = $ .... N m^{-1}

g Measure x_0 again. Has it changed? If so, how does this affect your value of k?

....

....

h Suppose you repeated the experiment with a stiffer spring. Draw a dotted line on the graph grid to show the expected result.

i In Table 1.4, list the advantages and disadvantages of using a newton-meter compared to a number of slotted masses.

Advantages	Disadvantages

Table 1.4: Advantages and disadvantages.

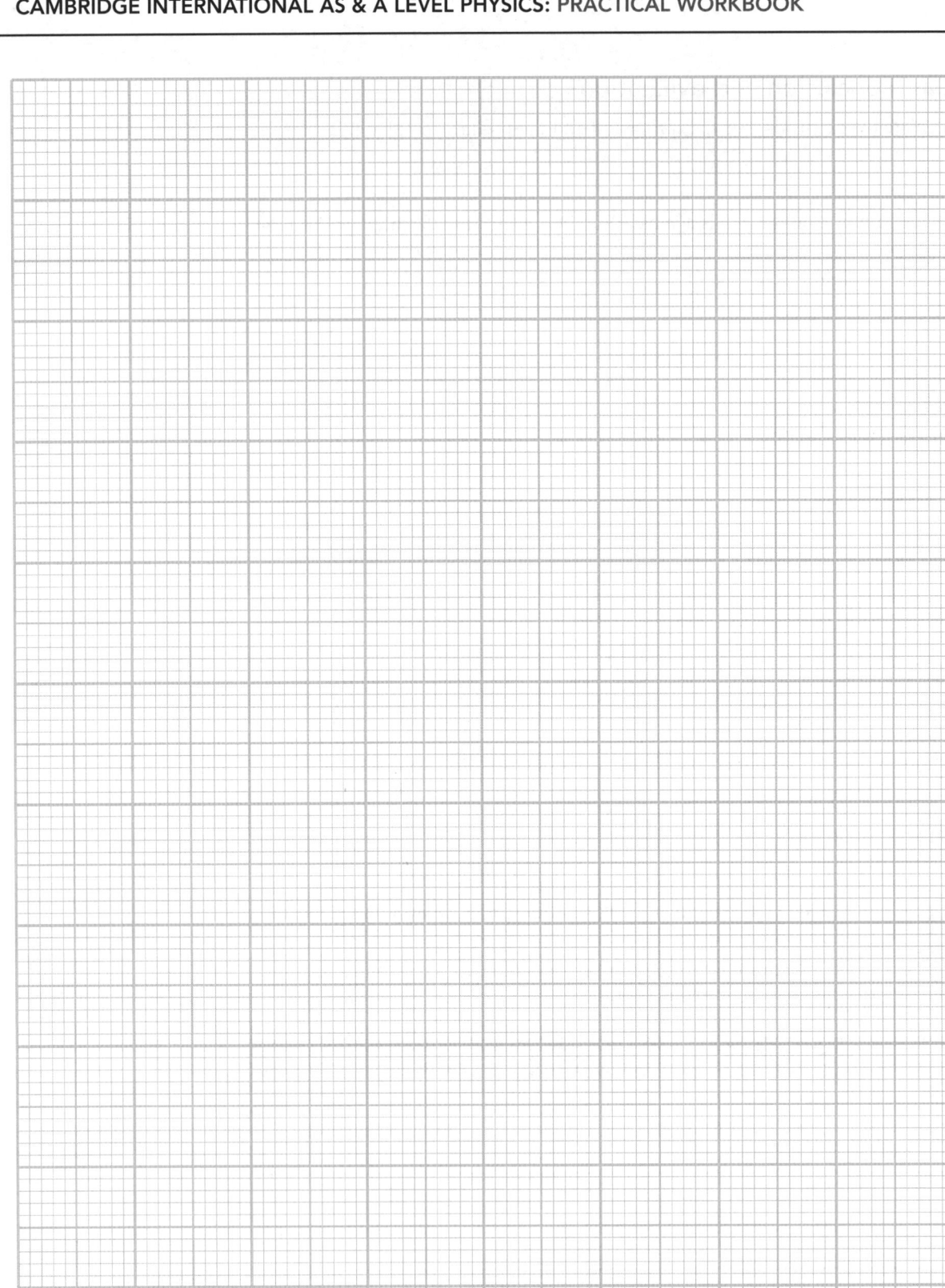

Practical investigation 1.3: Determining the resistance of a metal wire

The **resistance** of a resistor is defined by:

$$\frac{\text{potential difference across the resistor}}{\text{current in the resistor}}$$

or, expressed in symbols:

$$R = \frac{V}{I}$$

The standard unit for resistance in the SI system of units is the ohm (Ω).

> **KEY EQUATION**
>
> resistance $R = \dfrac{V}{I}$

> **YOU WILL NEED**
>
> **Equipment:**
>
> • 1.5 V cell • connecting leads • crocodile clips • power supply • two digital multimeters • rheostat • metre rule • switch.
>
> **Access to:**
>
> • reel of wire • scissors • adhesive tape • wire cutters • micrometer.

Safety considerations

* Make sure you have read the Safety advice at the beginning of this book and listen to any advice from your teacher before carrying out this investigation.

* There are no other specific safety issues with this investigation.

Part 1: Using digital multimeters

Method

1   Switch on one of the multimeters. When the dial is moved from the OFF position there are several possible ranges.

These could include:

* direct voltage
* alternating voltage
* direct current
* resistance.

Some of these ranges are shown in Table 1.5.

	Range from zero to:	Precision to the nearest:
direct voltage	600 V	1 V
	200 V	0.1 V
	20 V	0.01 V
	2000 mV (2 V)	1 mV (0.001 V)
	200 mV	0.1 mV
alternating voltage	600 V	1 V
	200 V	0.1 V
direct current	10 A	0.001 A
	200 mA	0.1 mA
	20 mA	0.01 mA
	2000 µA	1 µA
	200 µA	0.1 µA
resistance	2000 kΩ	
	200 kΩ	
	20 kΩ	
	2000 Ω	
	200 Ω	

Table 1.5: Different ranges of multimeters.

Check each range on your multimeter. They should all read zero. You can check the precision by noting where the decimal point is. If you have different ranges to those shown in Table 1.5, add them to the empty rows in this table.

The resistance ranges will all read '1'. This does *not* mean there is a reading of 1 Ω. It means the resistance that is being measured is off the top of the scale. Since no resistor is attached between the terminals of the multimeter, it is measuring a resistance of infinity on all the scales.

2 Connect the multimeter to the cell. If the reading is negative, reverse the connections to the meter.

   i   Go through the scales.

ii   For each scale, record the reading on the multimeter in Table 1.6.

Scale	Reading
600 V	
200 V	
20 V	
2000 mV	
200 mV	

Table 1.6: Results table.

3   Choose the most suitable scale and give reasons for your choice.

....

....

Part 2: Determining resistance from a single ammeter and voltmeter reading

Method

1   Use the wire cutters to cut a wire of length 110 cm.

2   Use the scissors to cut sufficient tape to attach the wire to the metre rule as shown in Figure 1.5.

Figure 1.5: Wire attached with tape to ends on a metre rule.

3   Connect the circuit as shown in Figure 1.6.

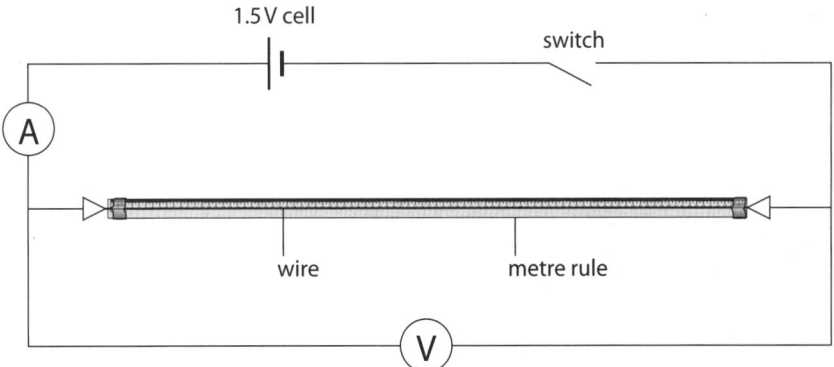

Figure 1.6: Circuit diagram for Part 2.

i   How many connecting leads do you need? ....

ii   How many crocodile clips do you need? ....

4   Switch both meters to suitable scales and record the readings in Table 1.7 in the Results section.

Results

Voltmeter reading V / V	Ammeter reading I / mA	I / A

Table 1.7: Results table.

Analysis, conclusion and evaluation

a   Calculate *R*.

R = ... Ω

Part 3: Using a rheostat

Method

The rheostat has three terminals, A, B and C, as shown in Figure 1.7.

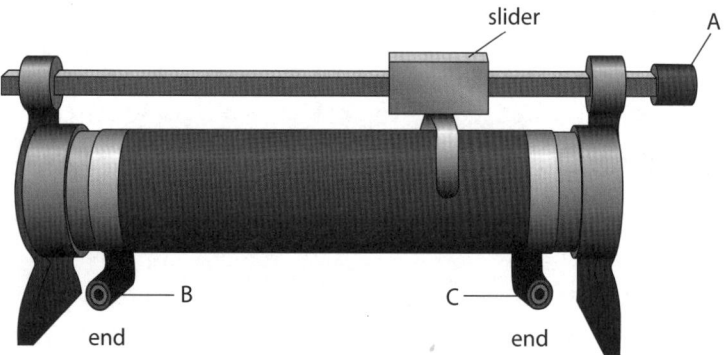

Figure 1.7: Rheostat showing three terminals, A, B and C.

1   Switch the multimeter to the 200 Ω range.

Connect the rheostat to the multimeter and complete Table 1.8.

Connections	Does the resistance reading change when the slider is moved?
A and B	yes / no
B and C	yes / no
A and C	yes / no

Table 1.8: Results table.

Part 4: Determining resistance using a graph

Method

1 Connect the rheostat into the circuit as shown in Figure 1.8.

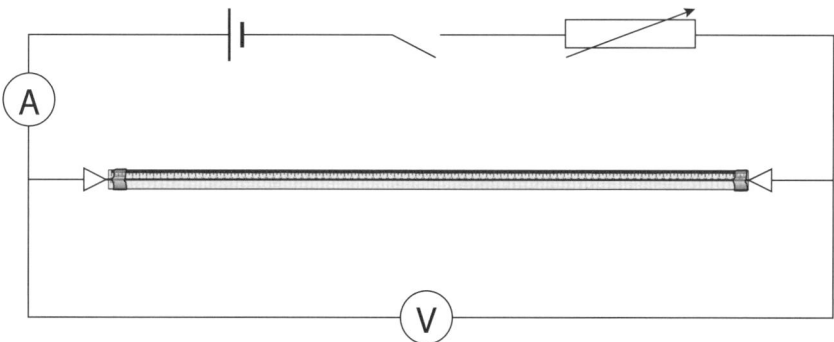

Figure 1.8: Circuit diagram for Part 4.

2 Move the slider on the rheostat and take a series of readings of V and I.

   Record these readings in Table 1.9 in the Results section.

3 Open the switch between readings to prevent discharging the battery.

Results

V / V	I / A

Table 1.9: Results table.

Analysis, conclusion and evaluation

a Plot a graph of I on the y-axis against V on the x-axis using the graph grid on the next page.

b Draw the straight line of best fit.

c Determine the gradient of this line.

Gradient = ...

d Determine R using $R = \dfrac{1}{\text{gradient}}$

R = ... Ω

e How could you use the rheostat to take a wide range of equally spaced readings?

....

....

Part 5: Using a micrometer

Resistance per unit length (resistance of 100 cm length of wire) depends on the diameter of the wire. Table 1.10 shows the properties of some wires A, B, C, D and E.

Wire	Diameter / mm	Resistance per unit length / Ω m^{-1}
A	0.38	4.4
B	0.27	8.3
C	0.19	16.8
D	0.15	27.0
E	0.10	60.0

Table 1.10: Properties of wires A, B, C, D and E.

Method

1 Use the micrometer to measure the diameter of your wire.

Diameter = ... mm

Analysis, conclusion and evaluation

a Use the data in Table 1.10 and your value of R to identify the most similar wire.

....

b Give a reason(s) for your choice.

....

....

c Theory suggests that the graph line in Part 4 should go through the point (0, 0).

Suppose you repeated the experiment with a wire of smaller diameter. Draw a dotted line on the graph grid in Part 4 to show the expected result.

Practical investigation 1.4: Determining the average speed of a cylinder rolling down a ramp

The **average speed** of an object is defined by:

$$\text{speed} = \frac{\text{distance travelled}}{\text{time taken}}$$

or, expressed in symbols:

$$v = \frac{d}{t}$$

The standard unit for speed in the SI system of units is m s^{-1}.

KEY EQUATION

average speed $v = \dfrac{d}{t}$

YOU WILL NEED

Equipment:

- cylinder • wooden board • stand • boss • clamp • metre rule • protractor
- stopwatch • book or pencil case to act as a barrier at the bottom of the ramp.

Safety considerations

- Make sure you have read the Safety advice at the beginning of this book and listen to any advice from your teacher before carrying out this investigation.

- Use a book or a pencil case to stop the cylinder after it has reached the bottom of the wooden board.

Part 1: Investigating reaction time

Method

1 Set your stopwatch to zero.

2 Switch the stopwatch on and off as quickly as you can and record the reading.

3 Repeat this reading twice more and record the three values in Table 1.11 in the Results section.

Results

t_1 / s	t_2 / s	t_3 / s

Table 1.11: Results table.

Analysis, conclusion and evaluation

a Calculate the mean value of t.

$t = \ldots\ldots\ldots\ldots$ s

b Figure 1.9 shows a reading of 1.44 seconds on a stopwatch.

Figure 1.9: Digital display reading 0:01(44).

Use your result in Table 1.11 to calculate the **percentage uncertainty** in the reading in Figure 1.9. You may assume that the absolute uncertainty in the reading on the stopwatch is the same as the absolute uncertainty in your readings in Table 1.1.

Percentage uncertainty = $\ldots\ldots\ldots\ldots$%

KEY WORDS

percentage uncertainty: the absolute uncertainty as a fraction of the measured value

---

KEY EQUATION

$$\text{percentage uncertainty} = \frac{\text{uncertainty}}{\text{mean value}} \times 100\%$$

Part 2: Determining average speed

Method

1 Set up the apparatus as shown in Figure 1.10.

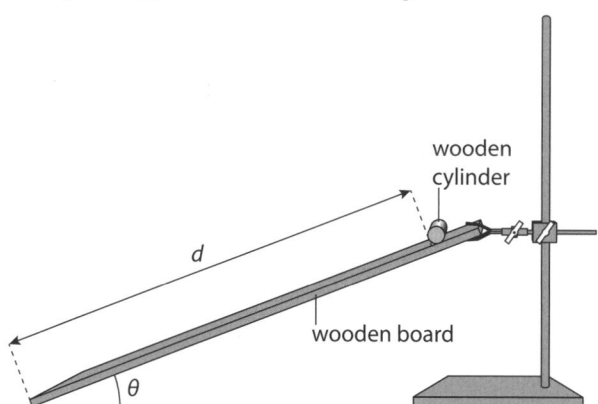

Figure 1.10: Wooden cylinder on sloping board.

2 Place the cylinder near the top of the wooden board.

Measure the distance d that the cylinder will travel down the wooden board when it is released. Write this value in the Results section.

3 Release the cylinder and measure the time t_1 for the cylinder to travel the distance d down the slope.

4 Repeat this reading and record the three values in Table 1.12 in the Results section.

Results

$d =$ .... cm

t_1 / s	t_2 / s	t_3 / s

Table 1.12: Results table.

Analysis, conclusion and evaluation

a Calculate the mean value of t from your results in Table 1.12.

Mean value of $t =$ .... s

b Calculate the average speed v.

$v =$ .... cm s^{-1}

Part 3: Investigating how the average speed depends on the angle of the plane

Method

1 Use the protractor to measure the angle θ between the plane and the bench as shown in Figure 1.10.

2 Take a series of readings of θ and t.

Record your data in Table 1.13 in the Results section.

Results

	t / s			
θ / °	1st value	2nd value	3rd value	Mean

Table 1.13: Results table.

Analysis, conclusion and evaluation

a   Use Table 1.14 to record calculated values of sin θ, t sin θ and v.

sin θ	t sin θ / s	v / cm s^{-1}

Table 1.14: Results table.

b   Plot a graph of v on the y-axis against t sin θ on the x-axis using the graph grid on the next page.

c   Draw the straight line of best fit.

d   Determine the gradient of this line.

Gradient = ...

e   The relationship between v, t and θ is:

$$v = \left(\frac{gt}{3}\right)\sin\theta$$

where g is the acceleration of free fall.

Use your gradient to determine a value for g.

g = ... m s^{-2}

f The accepted value for g is 9.81 m s^{-2} (or 981 cm s^{-2}) and the theory predicts that the y-intercept is zero.

Does your value for g differ from the accepted value?

....

....

g Does your straight line of best fit go through (0, 0)?

....

....

h Are there any anomalous point(s) that you did not include in your straight line of best fit?

....

....

Limitations and improvements

Practical investigation 2.1: Thermal energy loss from water in a polystyrene cup

The **thermal energy** loss from hot water contained in a polystyrene cup depends on several factors.

One of these factors is the mass of water in the cup. You are going to measure the time taken for the temperature of different masses of water to drop between two fixed temperatures.

KEY WORDS

thermal energy: energy transferred from one object to another because of a temperature difference; another term for heat energy

YOU WILL NEED

Equipment:
• long-stem thermometer: −10 °C to 110 °C × 1 °C • 200 cm³ polystyrene cup • stopwatch • stirrer • paper towel.

Access to:
• electric kettle or other means to heat water to boiling safely • top-pan balance • jug of cold water • waterproof pen.

Safety considerations

- Make sure you have read the Safety advice at the beginning of this book and listen to any advice from your teacher before carrying out this investigation.

- Take care when pouring hot water into the cup or emptying the hot water from the cup into a sink or container.

- When the thermometer is not in use, place it on the paper towel so that it does not fall onto the floor.

Method

1   Determine the mass of the cup using the top-pan balance. Record this value in the Results section.

2   Make three marks on the *inside* of the polystyrene cup at the positions shown in Figure 2.1.

3   Pour cold water into the cup until it reaches the bottom line.

4   Determine the mass of the cup and water. Record this value in Table 2.1 in the Results section.

5   Repeat steps **3** and **4** for the middle line and the top line.

6   Empty the water from the cup.

7   Pour boiling water into the cup until it reaches the bottom line.

8   Place the thermometer and stirrer in the cup.

9   Start the stopwatch when the temperature of the water in the cup is 85 °C.

10   Stop the stopwatch when the temperature of the water in the cup is 80 °C.

11   Record the time *t* in Table 2.1.

12   Repeat steps **6**, **7**, **8**, **9**, **10** and **11** for the middle line and the top line.

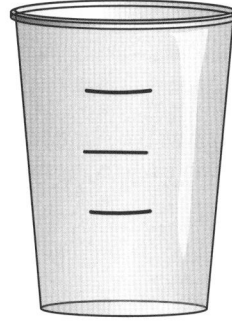

Figure 2.1: Polystyrene cup.

Results

Mass of cup = ... g

Mark on cup	Mass of cup and water / g	m / g	t / s	
bottom				
middle				
top				

Table 2.1: Results table.

Analysis, conclusion and evaluation

a   Calculate the mass of water m using:

m = mass of cup and water – mass of cup

and record these values in Table 2.1.

b   How does t vary with m?

....

....

c   Why was the same starting temperature (85 °C) and temperature change (5 °C) used in each experiment?

....

....

d   There is a spare column in Table 2.1.

Write in the spare column estimated times if a starting temperature of 80 °C had been used for the same temperature change (5 °C).

Suggest a reason for your estimated times.

....

....

....

e   In Table 2.2 there are some suggested limitations and improvements.

A and **B** require additional apparatus so it is appropriate to suggest the improvements.

It is appropriate to suggest that **C** and **D** are limitations. However, the improvements could have been achieved with the existing apparatus.

Suggest one more valid limitation and improvement and add these to row **E** in Table 2.2.

	Limitation	Improvement
A	The rate of thermal energy loss depends on the area of the water surface exposed to the air. In each experiment the area was different.	Use a cup that has straight vertical sides.
B	It was difficult to read the thermometer, stir the water and hold the stopwatch at the same time.	Use a stand and clamp to hold the thermometer.
C	After the first experiment the cup was warmed up so this may have affected the remaining results.	Rinse the cup with cold water after each experiment.

	Limitation	Improvement
D	If room temperature had changed during the experiment it would have affected the results.	Check room temperature before and after each experiment.
E	....	....

Table 2.2: Limitations and improvements.

Practical investigation 2.2: Loaded rubber band

When a material is subject to a force it undergoes **tensile** deformation. You are going to investigate the relationship between the force exerted on a rubber band and its extension.

YOU WILL NEED

Equipment:

• two stands • two bosses • two clamps • G-clamp • 100 g mass hanger • four 100 g slotted masses • protractor • metre rule • rubber band with approximate cross-section 2 mm × 1 mm and approximate circumference 20 cm.

Safety considerations

• Make sure you have read the Safety advice at the beginning of this book and listen to any advice from your teacher before carrying out this investigation.

• Take care when moving the stand. It may topple when the rubber band is stretched.

• Do not try to extend the rubber band as much as possible. This could break the rubber band causing the masses to fall to the bench or the floor.

Part 1: Suspending the rubber band from two rods

Method

1   Set up the apparatus as shown in Figure 2.2. The distance x between the rods of the clamps should be approximately 10 cm.

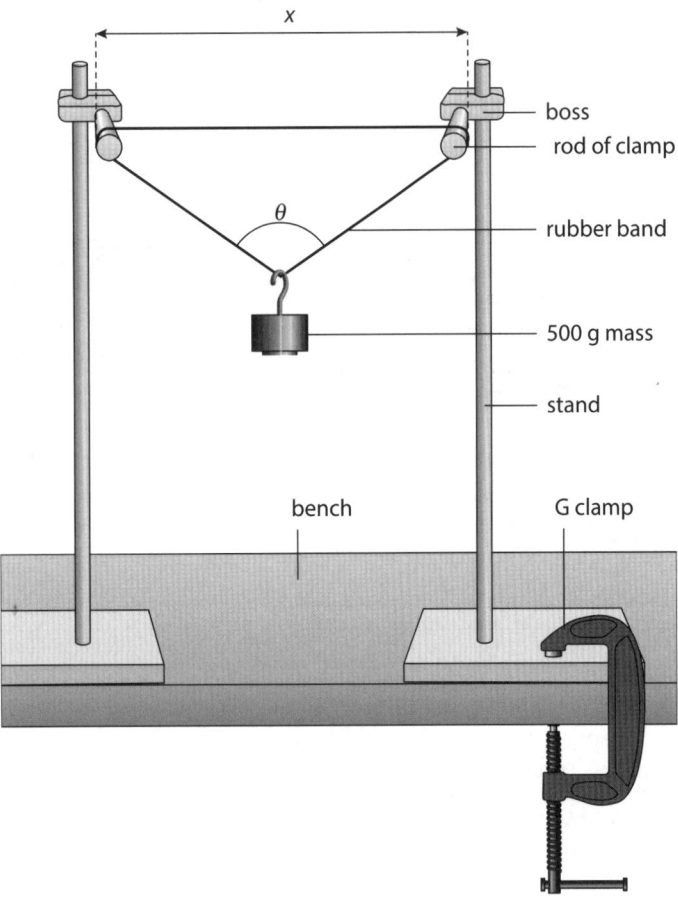

Figure 2.2: Rubber band, with weight, suspended from two rods.

2   Record your value of x in Table 2.3 in the Results section.

3   Measure the angle θ and record this value in Table 2.3.

4   Move the left-hand stand to the left and take a series of readings of x and θ.

   Record your data in Table 2.3.

5   After you have taken all your readings, remove the rubber band from the clamps.

   Measure its total length (circumference) C.

   Record this value under Table 2.3.

Results

x / cm	θ / °	$\sin\left(\dfrac{\theta}{2}\right)$	L / cm	e / cm

Table 2.3: Results table.

$C =$ .... cm

Analysis, conclusion and evaluation

a The total length L of the extended rubber band is given by:

$$L = x + \frac{x}{\sin\left(\dfrac{\theta}{2}\right)}$$

Calculate values of $\sin\left(\dfrac{\theta}{2}\right)$ and L and record your values in Table 2.3.

b How does L vary with x?

....

....

c Calculate the extension e of the rubber band using:

$$e = L - C$$

Record your values of e in Table 2.3.

Part 2: Suspending the rubber band from one rod

Method

1   Set up the apparatus as shown in Figure 2.3. The distance between the top of the rod of the clamp and the hook of the mass hanger is R.

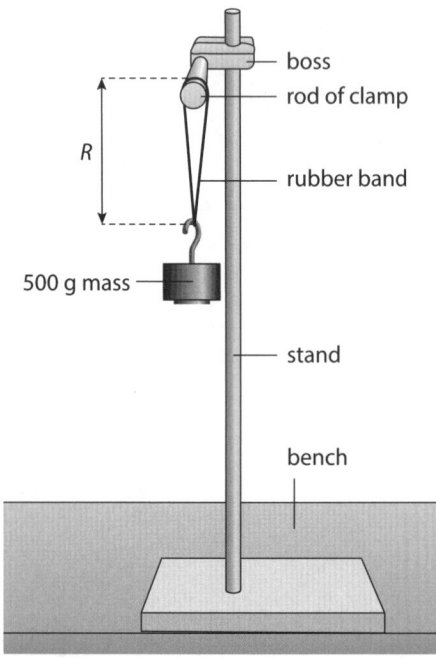

Figure 2.3: Rubber band, with weight, suspended from one rod.

2   Measure and record R.

$R = $ .... cm

Analysis, conclusion and evaluation

a   Calculate the extension e of the rubber band using:
$e = 2R - C$

Extension = ... cm

b   Look at the data in Table 2.3 and estimate the value of x that would result in the same extension that you determined in Part 1.

Estimated value of $x = $ .... cm

c   In Table 2.4 there are three suggested limitations and improvements, but only **two** of them are acceptable.

In **A** and **B** either a different technique or the use of extra apparatus is suggested.

In **C** the suggestion could have been performed using the existing apparatus, so this is **not** a good suggestion.

Write two more limitations and improvements in rows **D** and **E** in Table 2.4.

	Limitation	Improvement
A	It was difficult to measure θ because the hook of the mass hanger was in the way.	Set up a card behind the rubber band. Draw lines on the card which are parallel to the rubber band. Measure θ on the card.
B	It was difficult to measure θ because the hook of the mass hanger was in the way.	Tie a thin string loop around the rubber band from which to suspend the mass.
C	It was difficult to measure θ because the hook of the mass hanger was in the way.	Determine $\dfrac{\theta}{2}$ by measuring lengths and using trigonometry.
D	....	....
E	....	....

Table 2.4: Limitations and improvements.

Practical investigation 2.3: Balanced metre rule

When a system is in **equilibrium** there is no resultant force and no resultant torque. You are going to use a balanced metre rule to determine an unknown mass.

YOU WILL NEED

Equipment:

• stand • boss • clamp • metre rule • loop of thick string of circumference 20 cm • 50 g slotted mass • three 10 g slotted masses • mass M of modelling clay • small triangular pivot.

Safety considerations

- Make sure you have read the Safety advice at the beginning of this book and listen to any advice from your teacher before carrying out this investigation.

- If the masses move off the rule when it slides through the string loop, move the boss closer to the bench.

Method

1 Set up the apparatus as shown in Figure 2.4. The metre rule should be balanced with the centres of mass M and the 50 g mass each positioned 2 cm from each end of the rule. The distance between the centre of the 50 g mass and the string is y.

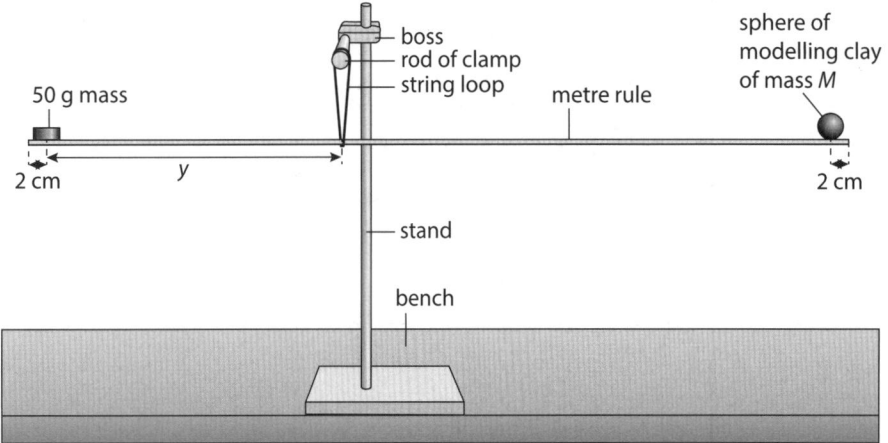

Figure 2.4: Metre rule balanced with weights and sphere of modelling clay.

Record this value of mass m (50 g) and the value of y in Table 2.5.

2 Change the mass m and slide the metre rule through the string loop until it is balanced again. Record the new values of m and y in Table 2.5.

3 Repeat step 2 with different values of m and record your data in Table 2.5.

> **TIP**
>
> The masses provided allow you to have seven different values of m.

Results

m / g	y / cm

Table 2.5: Results table.

Analysis, conclusion and evaluation

a   How does *y* vary with *m*?

....

....

b   Plot a graph of *y* on the *y*-axis against *m* on the *x*-axis using the graph grid on the next page.

c   Draw the curve of best fit through your points.

d   When *m* and *M* are the same, the metre rule should be balanced when the string is at the mid-point of the metre rule (i.e. $y = 48$ cm). Use your graph to determine *M*.

e   In Table 2.6 there is one suggested limitation and improvement in row **A**. Write two more limitations and improvements in rows **B** and **C** in this table.

	Limitation	Improvement
A	It was difficult to read the scale on the metre rule because the string was too thick.	Use thinner string.
B	....	....
C	....	....

Table 2.6: Limitations and improvements.

> **TIP**
>
> Try using a small triangular pivot: is it an improvement? If not, why not?

Kinematics and dynamics

Practical investigation 3.1:
Acceleration of connected masses

If a string over a pulley has a mass attached to each end, any difference in the masses will cause the system to accelerate. In this investigation, part of one mass is transferred to the other so that the mass difference is changed but the total mass is constant.

YOU WILL NEED

Equipment:
• pulley wheel to clamp to edge of bench • thin string • two mass hangers, each with a total mass of 500 g • two paper clips • 20 steel washers (steel rings) • stopwatch • metre rule • thick cardboard mat.

Safety considerations

- Make sure you have read the Safety advice at the beginning of this book and listen to any advice from your teacher before carrying out this investigation.

- One of the masses will hit the cardboard mat on the floor. You must keep your feet away from this area.

Method

1 Set up the apparatus as shown in Figure 3.1.

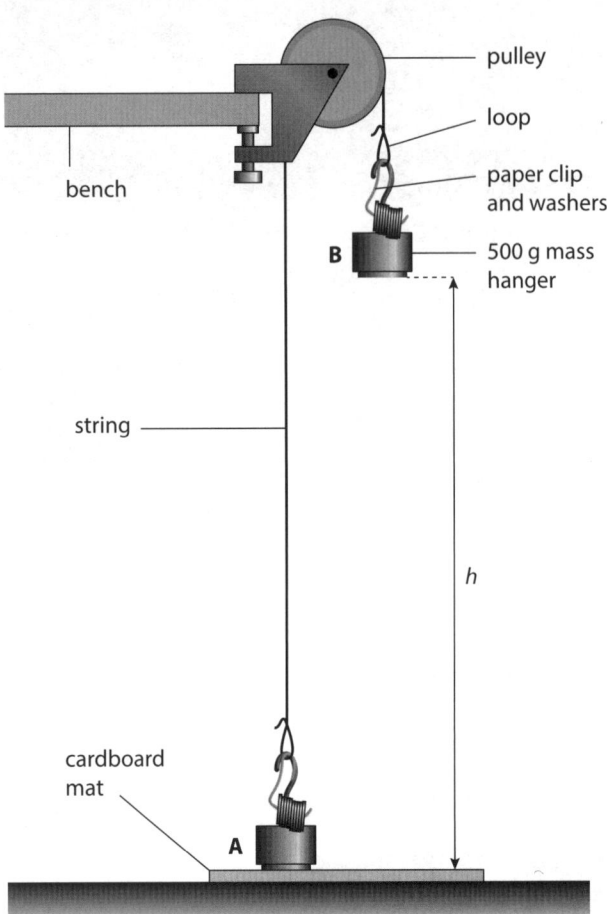

Figure 3.1: Weights, paper clips and string over pulley.

2 Tie a loop at each end of the string.

3 Open out each paper clip into a hook and thread 10 washers onto it.

4 Hook a 500 g mass and a paper clip onto each loop of string.

5 Pull mass hanger A down until it is touching the cardboard mat and measure the height h of mass hanger B above the mat. Record h in the Results section.

6 Move washers, one at a time, from A to B until B starts to move down steadily to the mat.

7 Record the difference n between the number of washers on A and the number of washers on B in Table 3.1 in the Results section.

8 Lift B up until A is just touching the mat, then release it and measure the time t for it to reach the mat. Record the value in Table 3.1.

9 Move more washers from A to B and repeat steps **7** and **8**. Repeat until you have six sets of values for n and t in Table 3.1.

TIP
Remember to record metre rule measurements to the nearest mm.

TIP
Measure t several times for each n and record all the values.

Results

$h =$ .... cm

n	t / s				a / cm s^{-2}
	first	second	third	mean	

Table 3.1: Results table.

Analysis, conclusion and evaluation

a For each row in Table 3.1 calculate the mean value of t.

b For each row in Table 3.1 calculate the **acceleration** a using:

$$a = \frac{2h}{t^2}$$

c Use the grid to plot a graph of a (on the vertical axis) against n (on the horizontal axis).

d Draw the straight line of best fit through the points.

e Determine the gradient and intercept of the line.

Gradient = ...   Intercept = ...

f The theory for this experiment is based on:

mass × acceleration = force

and since the total mass is constant this gives:

a is proportional to n

In practice, the graph does *not* pass through the origin. Explain why the first transfers of washers do *not* produce an acceleration.

....

....

KEY WORD

acceleration: the rate of change of velocity of an object

KEY EQUATION

$s = ut + \frac{1}{2}at^2$

TIP

Refer to the Skills chapter for advice on best-fit lines.

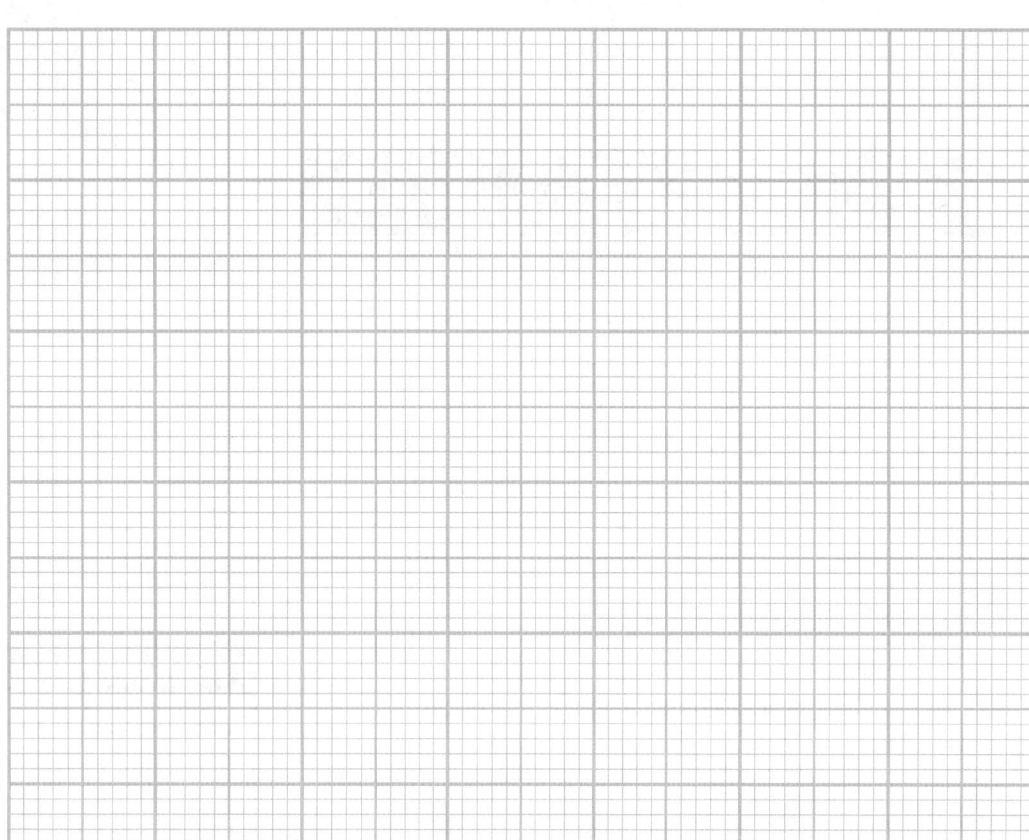

Practical investigation 3.2:
Energy and amplitude of a pendulum

The arrangement of apparatus enables the bob of a pendulum to be given the same amount of energy each time in repeated tests. The investigation looks at how the pendulum **amplitude** varies with its length.

YOU WILL NEED

Equipment:

- table tennis ball with length of thread attached • stand, boss and clamp
- glass marble in small tray • inclined pipe held in stand • rectangular block
- metre rule.

Safety considerations

- Make sure you have read the Safety advice at the beginning of this book and listen to any advice from your teacher before carrying out this investigation.

- There are no special safety issues with this experiment.

Method

1 Set up the apparatus as shown in Figure 3.2. The angle and height of the pipe have been set for you. Do *not* adjust the pipe.

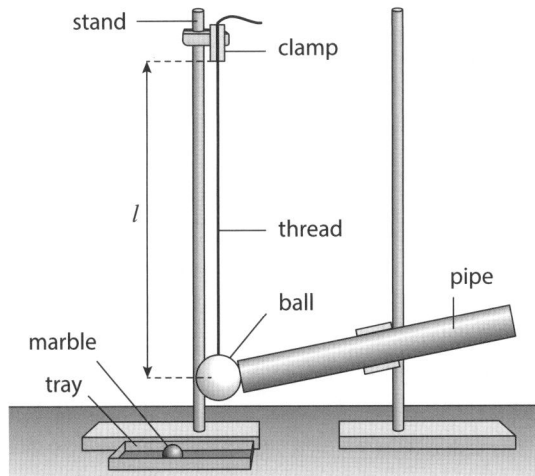

Figure 3.2: Sloping pipe, ball on string and a marble.

2 Adjust the thread in the clamp and the clamp height until the length *l* to the centre of the ball is approximately 50 cm and the ball is just touching the end of the pipe.

3 Measure *l* and record the value in Table 3.2.

4 Place the marble in the top of the pipe so that it rolls down and hits the ball. The ball will swing out a horizontal distance *d*, as shown in Figure 3.3. Repeat this several times, moving the rectangular block closer until the ball just reaches it as it swings.

> **TIP**
>
> This uses a 'trial and error' method to find *d*.

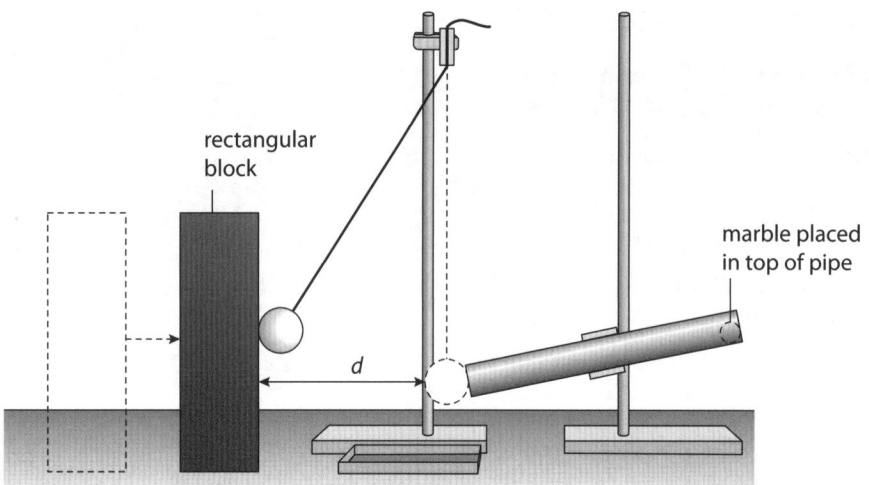

Figure 3.3: As Figure 3.2, but with the marble being inserted in the tube and the block placed at the point that the ball just reaches.

5 Measure the amplitude d and record the value in Table 3.2.

6 Reduce l by about 5 cm by moving the thread in the clamp.

7 Lower the clamp holding the thread so that the ball is just touching the end of the pipe again, then repeat steps **3**, **4** and **5**. *Do not adjust the pipe.*

8 Repeat steps **6** and **7** until you have six sets of values of l and d in Table 3.2.

Results

l / cm	d / cm	d² / cm²

Table 3.2: Results table.

Analysis, conclusion and evaluation

a Calculate the values of d^2 and add them to Table 3.2.

b Use the grid to plot a graph of d^2 (on the vertical axis) against l (on the horizontal axis).

> **TIP**
>
> For step 5, you may feel that there is some uncertainty in d, but as you are measuring with a metre rule you should record your value to the nearest mm.

> **TIP**
>
> Choose scales so that the points use most of the grid (refer to the Skills chapter).

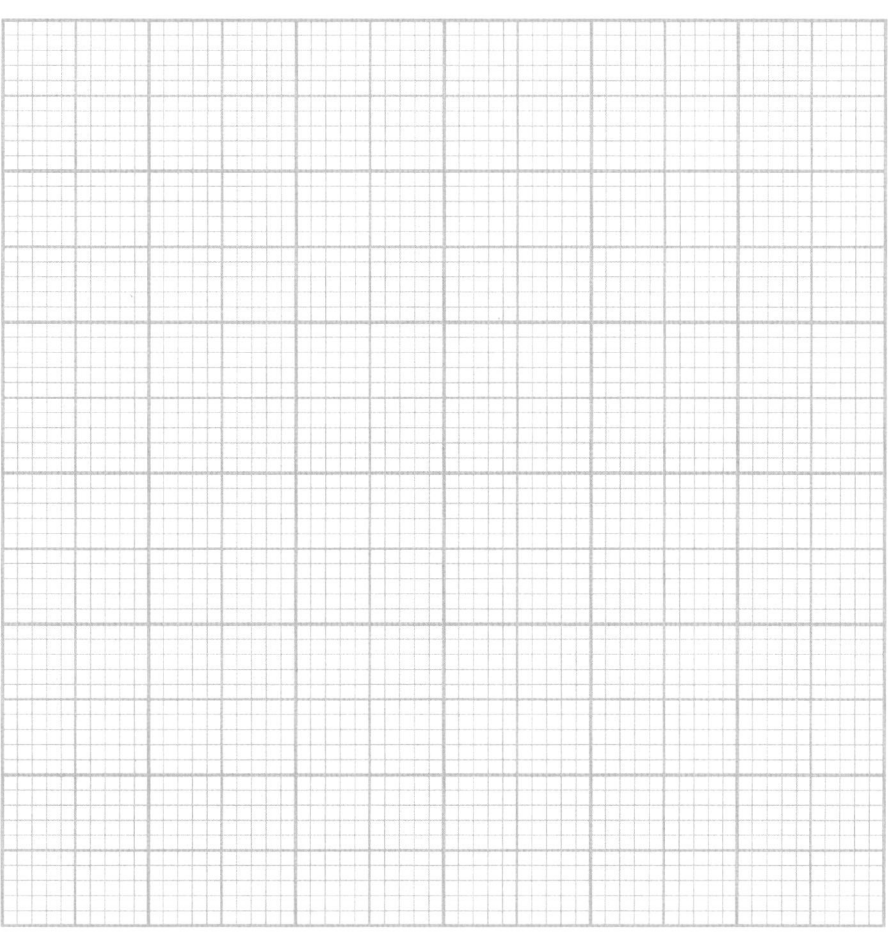

c   Draw the line of best fit through the points.

d   Determine the gradient and intercept of the line.

Gradient = ...   Intercept = ...

e   It is suggested that l and d are related by the equation $d^2 = Al + B$, where A and B are constants. Use your answers from part **d** to determine the values of A and B. Give suitable units.

A = ... B = ...

f   This experiment relies on the marble hitting the ball at the same speed each time: this impact speed is a controlled variable. Given that the length and slope of the tube are constant, how could this impact speed vary from one test to the next?

....

....

Practical investigation 3.3: Range of a projectile

In this practical investigation you will investigate how far a ball travels when it is launched horizontally at different heights above a tray of sand. The data is used to calculate the horizontal launch **velocity**.

KEY WORD

velocity: an object's speed in a particular direction, or the rate of change of an object's displacement; it is a vector quantity

YOU WILL NEED

Equipment:

• curved tube fixed to a cardboard rectangle • steel ball (ball bearing) in a small container • tray of sand • pencil • stand, boss and clamp • set square • 30 cm ruler • metre rule.

Safety considerations

- Make sure you have read the Safety advice at the beginning of this book and listen to any advice from your teacher before carrying out this investigation.
- There are no special safety issues with this experiment.

Method

1 Set up the apparatus as shown in Figure 3.4.

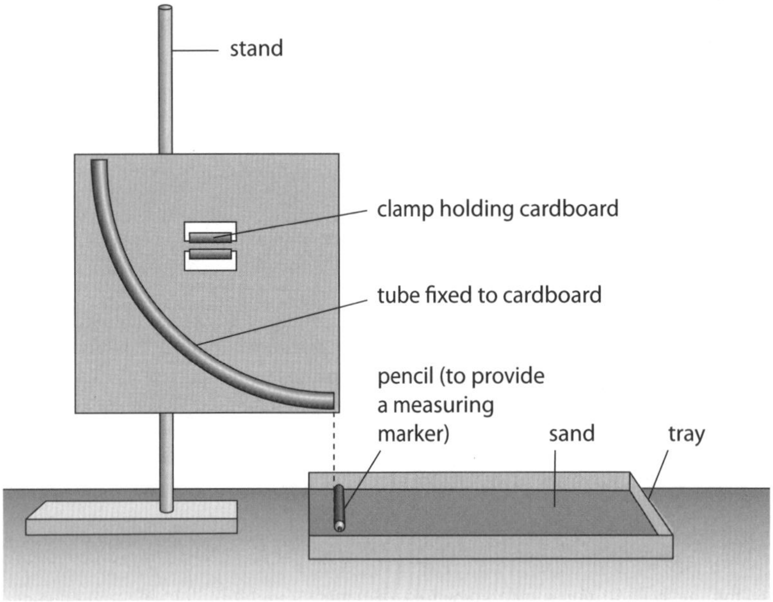

Figure 3.4: Curved tube fixed on cardboard sheet.

The bottom edge of the cardboard should be parallel to the bench.

The horizontal pencil should be pushed into the sand so that it is vertically below the end of the tube and parallel to the end of the tray, as shown on the diagram.

TIP

In two places, measure the height of the cardboard above the bench. The measurements should be the same.

2   Measure the height h of the tube above the sand, as shown in Figure 3.5, and record it in Table 3.3 in the Results section.

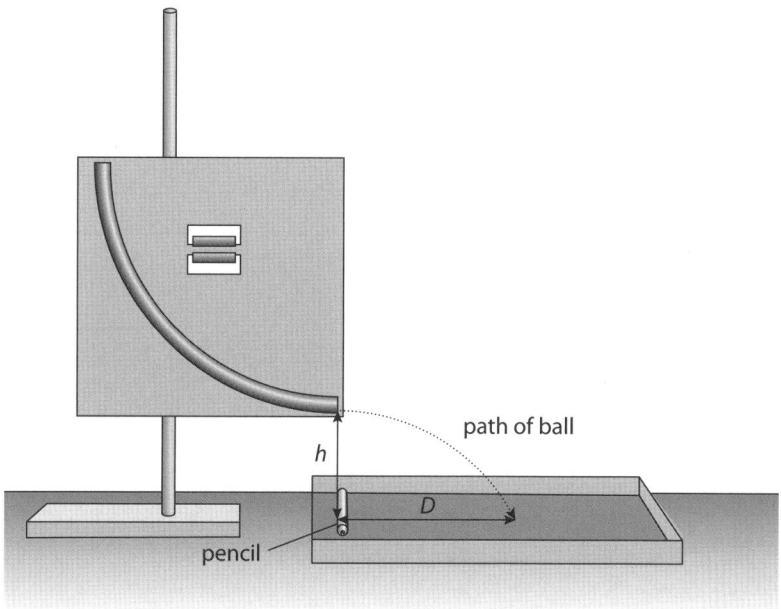

path of ball

h

D

pencil

Figure 3.5: As Figure 3.4, but with path of ball shown.

3   Put the steel ball into the top of the tube so that it rolls out and falls onto the sand.

4   Measure the distance D from the ball landing position to the pencil, as shown in Figure 3.5, and record it in Table 3.3.

5   Pick up the ball and smooth the sand with the set square.

6   Repeat steps **3** and **4** several times, recording the results in Table 3.3 and calculating the mean value of D.

7   Change h and repeat steps **2–6** until you have six sets of values of h and mean D in Table 3.3.

> **TIP**
>
> After each change of h check that the end of the tube is horizontal and is vertically above the pencil.

Results

h / cm	D / cm					D^2 / cm^2
	1	2	3	4	mean	

Table 3.3: Results table.

Analysis, conclusion and evaluation

a   Calculate the values for D^2 and enter them in Table 3.3.

b   Use the graph grid to plot a graph of D^2 (on the vertical axis) against h (on the horizontal axis).

c   Draw the line of best fit through the points.

d   Determine the gradient and intercept of the line.

Gradient = ...   Intercept = ...

e   The theory for the motion of a horizontal projectile suggests that the gradient of the graph is equal to $\dfrac{2v^2}{g}$, where v is the horizontal velocity and $g = 9.81$ m s^{-2}.
Use your value for the gradient to calculate v. Include the unit.

$v = $ ....

TIP

Convert g to cm s^{-2} before calculating v.

Practical investigation 3.4: Terminal velocity of a ball falling through water in a tube

Figure 3.6 shows part of a tube filled with water. As a ball falls through the tube, water has to move from below the ball to above it, flowing through the gap between the ball and the walls of the tube.

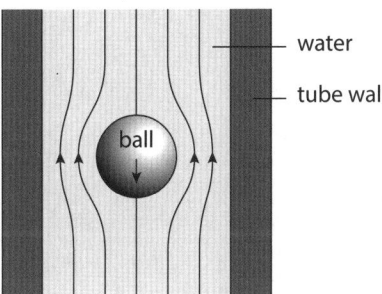

Figure 3.6: Flow of water around a ball being dropped in a tube.

In this investigation you will test how the drag force varies with the size of the gap between the ball and the tube.

YOU WILL NEED

Equipment:

• tall U-shaped plastic tube filled with water • short sample of the same plastic tube • two sizes of steel ball (five of each size) in a small tray • digital callipers • stopwatch • magnet • metre rule.

Safety considerations

- Make sure you have read the Safety advice at the beginning of this book and listen to any advice from your teacher before carrying out this investigation.

- There are no special safety issues in this investigation.

Method

1 The tall U-shaped plastic tube has been set up for you as shown in Figure 3.7. One leg of the tube has two marks on it. Measure the distance L between the upper mark and the lower mark. Record the value of L in the Results section.

> **TIP**
>
> Make sure your value for L matches the unit on the answer line.

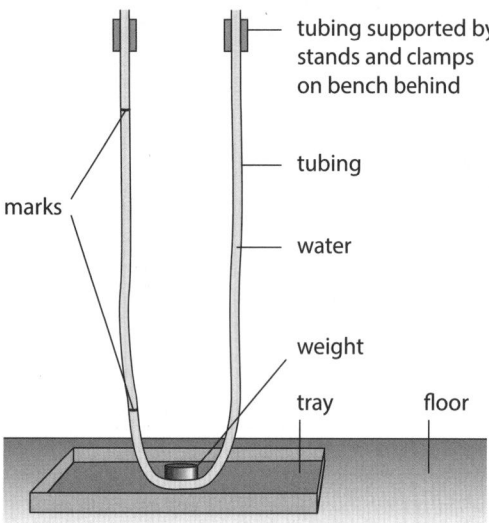

marks

tubing supported by stands and clamps on bench behind

tubing

water

weight

tray   floor

Figure 3.7: Suspended 'U'-shaped plastic tubing.

2 Measure the inside diameter D of the short sample of plastic tube and record the value in the Results section.

3 Measure the diameter d of one of the smaller steel balls and record the value in Table 3.4 in the Results section.

4 Drop one of the smaller steel balls into the tube and measure the time T it takes to fall from the upper mark to the lower mark. Record the value in Table 3.4.

> **TIP**
>
> Look at the advice on timing experiments in the Skills chapter.

5 Repeat step **4** for the rest of the smaller balls. If necessary, the balls can be lifted out of the tube using the magnet.

6 Repeat steps **3**, **4** and **5** using the larger balls.

Results

$L = $ .... m $D = $ .... mm

	d / mm	Values of T / s				
Smaller balls						
Larger balls						

Table 3.4: Results table.

Analysis, conclusion and evaluation

a   For each row in Table 3.4, calculate the mean value for T and record it in Table 3.5.

	T / s	v	A	k
Smaller balls				
Larger balls				

Table 3.5: Results table.

b   For each row in Table 3.4, calculate the ball velocity v using:

$$v = \frac{L}{T}$$

Record the values in Table 3.5.

c   For each row in Table 3.4, calculate the area A of the gap between the ball and the walls of the tube using:

$$A = \frac{\pi(D^2 - d^2)}{4}$$

Record the values in Table 3.5.

d   Add the units for v and A in the headings in Table 3.5.

e   It is suggested that v and A are related by:

$$v = kA$$

where k is a constant.

For each row in Table 3.4, calculate the value of k and enter it in Table 3.5.

f   Calculate the percentage difference between the two values of k.

Percentage difference = ...%

g   Estimate the percentage uncertainty in your value of T for the smaller balls.

Percentage uncertainty = ...%

> **TIP**
>
> When calculating the percentage, do *not* use the smallest division on the stopwatch; use an estimate of the uncertainty due to human reaction time (e.g. 0.2 s).

h Explain whether your answers in parts **f** and **g** support the relationship suggested in part **e**.

....

....

....

Forces, work and energy

Practical investigation 4.1: Effect of load position on beam supports

The upward forces at each end of a simple beam bridge have to support the weight of the beam itself as well as the weight of any additional load on the beam. This practical investigates how one of these supporting reaction forces changes as a load moves along the beam.

YOU WILL NEED

Equipment:
- wooden strip with a string loop near each end • another string loop
- newton-meter • mass with a hook, labelled M • two stands and two bosses
- metre rule.

Safety considerations

- Make sure you have read the Safety advice at the beginning of this book and listen to any advice from your teacher before carrying out this investigation.

- There are no special safety issues in this investigation.

Method

1 Set up the apparatus as shown in Figure 4.1.

   M should be positioned so that the distance x is approximately 20 cm.

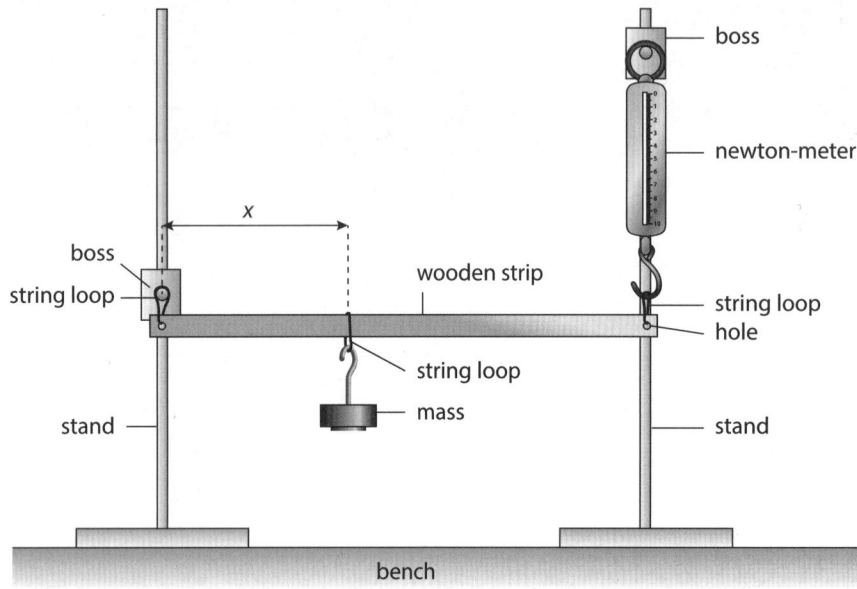

Figure 4.1: Wooden strip suspended, with weight and newton-meter.

2 Measure the distance L between the two holes in the wooden strip. Record the value of L in the Results section.

3 Adjust the heights of the bosses until the wooden strip is parallel to the bench.

4 If necessary, move the stands until the newton-meter is vertical.

5 Measure the distance x and the newton-meter reading F. Record the values in Table 4.1 in the Results section.

6 Change x and repeat steps **3**, **4** and **5** until you have six sets of values of x and F in Table 4.1.

Results

L = ... cm

x / cm	F / N

Table 4.1: Results table.

TIP
For step **3**, measure the height of the strip above the bench in two places. The measurements should be the same.

TIP
For step **6**, move the mass to give a wide range of x values, both above and below the initial value.

TIP
Remember to record metre rule measurements to the nearest mm.

Analysis, conclusion and evaluation

a Use the graph grid to plot a graph of F (on the vertical axis) against x (on the horizontal axis).

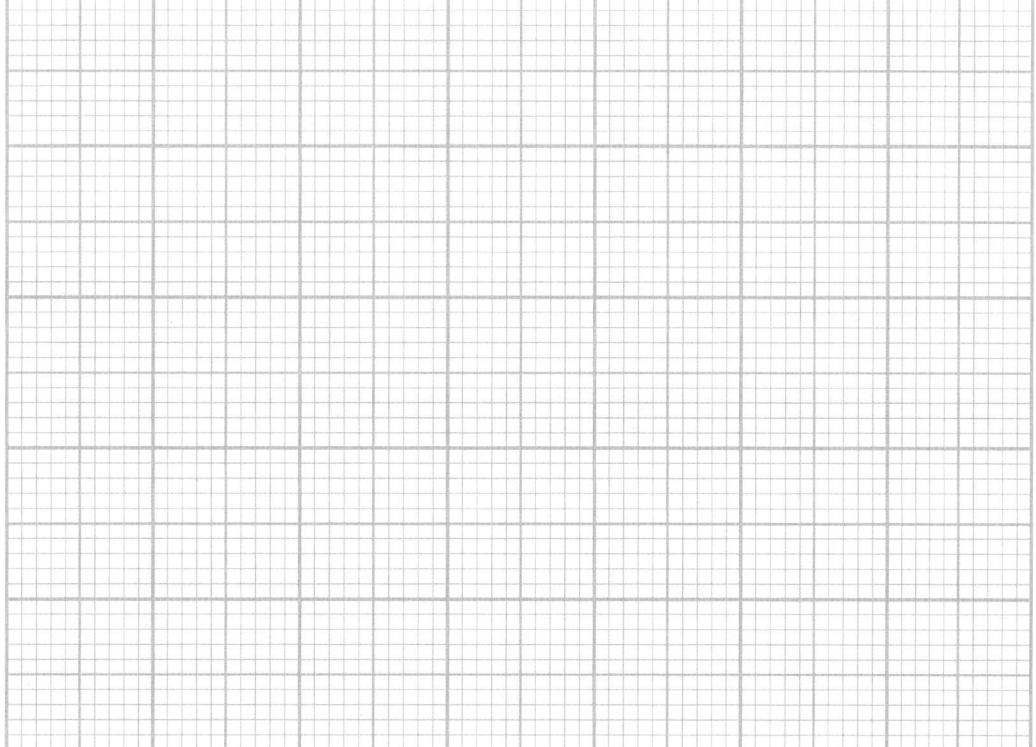

b Draw a straight line of best fit through the points.

c Determine the gradient and intercept of the line.

> **TIP**
>
> For part **b**, refer to the Skills chapter for advice on best-fit lines.

Gradient = ...   Intercept = ...

d The theoretical relationship between F and x is:

$$F = \frac{Wx}{L} + \frac{Z}{2}$$

where W is the weight of M and Z is the weight of the wooden strip.

Use your answers from part **c** to calculate the values of W and Z.

Include appropriate units.

> **TIP**
>
> You can use the fact that the unit for each term in the equation must be the same.

$W = $ .... $Z = $ ....

e If the beam was supported from below, what laboratory equipment could be used to measure the force at one of the supports?

....

Practical investigation 4.2: Determining the density of a metal sample

In this practical exercise you will use an **upthrust** method to determine the volume and density of an irregularly shaped metal sample. The forces are compared using their turning **moments**.

KEY WORDS

upthrust: the force upwards in a liquid caused by the difference in hydrostatic pressure in the liquid

moment: the moment of a force about a point is the product of the force and perpendicular distance from the line of action of the force to the point

YOU WILL NEED

Equipment:
• metal sample with loop of thread attached • metre rule • 100 g mass with loop of thread attached • knife-edge pivot • wooden block • beaker
• jug of water • paper towels.

Safety considerations

• Make sure you have read the Safety advice at the beginning of this book and listen to any advice from your teacher before carrying out this investigation.

• Mop up any spilled water to reduce the risk of slipping.

Method

1 Set up the apparatus as shown in Figure 4.2.

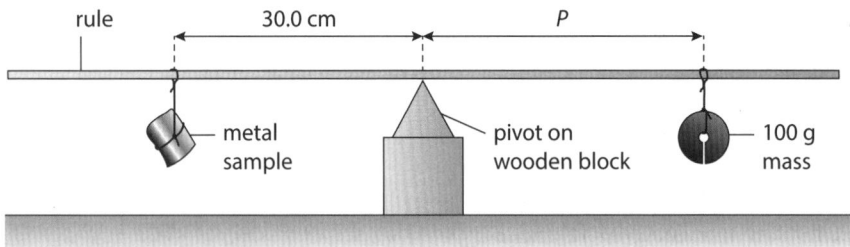

Figure 4.2: Rule balanced on a pivot with weight and metal sample.

i Position the metre rule so that the 50 cm mark is over the pivot.
ii Move the 100 g mass until the rule is balanced.
iii Measure P and record the value in the Results section.

2 Position the beaker of water as shown in Figure 4.3.

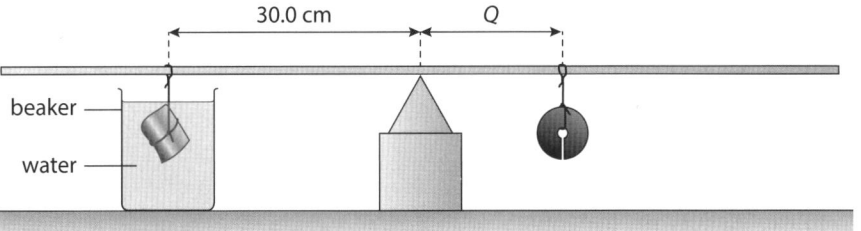

Figure 4.3: Similar to Figure 4.2, but with metal sample in beaker of water.

i Add water until the metal sample is completely immersed (but not touching the base of the beaker).

ii Move the 100 g mass until the rule is balanced.

iii Measure Q and record the value in the Results section.

Results

P = ... cm Q = ... cm

a Calculate the mass M of the metal sample using the relationship $100P = 30.0M$.

M = ... g

b Calculate the volume V of the metal sample using the relationship:

$$V = \frac{100(P-Q)}{30.0\rho_w}$$

where ρ_w is the density of water and is equal to 1.000 g cm^{-3}.

V = ... cm^3

c Calculate the density D of the metal sample using the relationship:

$$D = \frac{M}{V}$$

D = ... g cm^{-3}

d Express D in the alternative units of kg m^{-3}.

$D =$ .... kg m^{-3}

Analysis, conclusion and evaluation

e Look up density values to see if you can identify the metal in the sample.

....

f Describe one difficulty with using the apparatus and how it could increase the uncertainty in D.

....

....

Practical investigation 4.3: Equilibrium of a pivoted wooden strip

In this practical you will investigate the **equilibrium** angle of a wooden strip when pulled aside by a horizontal force. The force is provided by a spring, and the length of the spring depends on the magnitude of the force.

YOU WILL NEED

Equipment:

• wooden strip with a hole near one end and a string loop attached at its centre • spring with end loops • nail • two stands (one at least a metre tall) • two G-clamps • boss • plumb line • protractor • modelling clay • ruler and metre rule.

Safety considerations

• Make sure you have read the Safety advice at the beginning of this book and listen to any advice from your teacher before carrying out this investigation.

• The stands may tilt over so it is important that they are clamped to the bench.

Method

1   Set up the apparatus as shown in Figure 4.4.

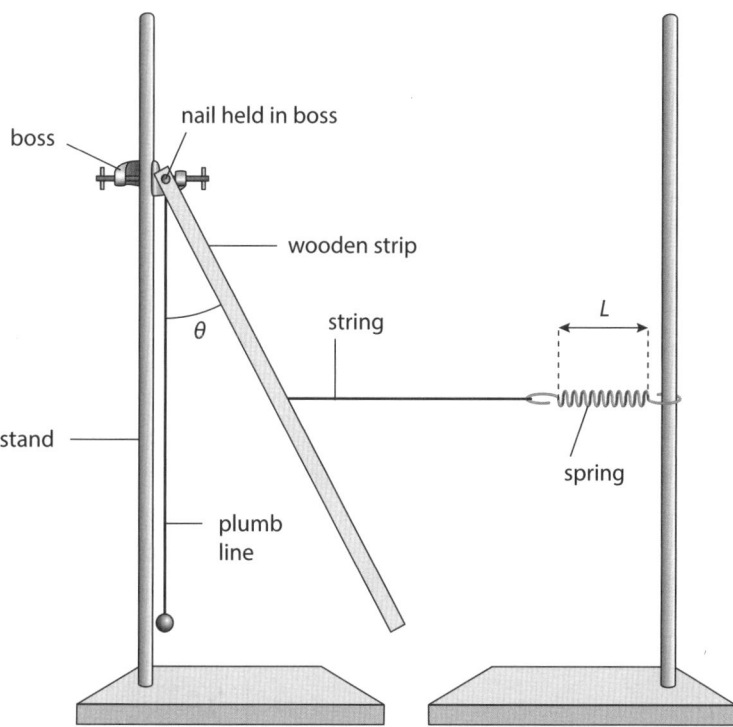

Figure 4.4: Wooden strip (pivoted) held at angle by string attached to spring.

i   Use the G-clamps to secure the stand bases to the bench.

ii   The nail should be held firmly in the boss.

iii Suspend the plumb line from the nail.

iv   The wooden strip should be able to pivot freely on the nail.

v   The initial value of θ should be approximately 30°.

2   Slide the spring up or down the stand until the string is horizontal.

3   Measure the angle θ that the wooden strip makes with the plumb line and record the value in Table 4.2 in the Results section.

4   Measure the length L of the coiled part of the spring and record the value in Table 4.2.

5   Change θ by moving the stand holding the spring and repeat steps **2**, **3** and **4** until you have six sets of values of θ and L in Table 4.2.

6   Add units to the headings in Table 4.2.

> **TIP**
>
> For step 2, measure the height of the string above the bench in two places. The measurements should be the same.

> **TIP**
>
> For step 6, since tan θ is a ratio it doesn't have a unit.

Results

θ /	L /	$\tan \theta$

Table 4.2: Results table.

Analysis, conclusion and evaluation

a Calculate the values for $\tan \theta$ and enter them in Table 4.2.

b Use the grid to plot a graph of $\tan \theta$ (on the vertical axis) against L (on the horizontal axis).

c Draw the line of best fit through the points.

d Determine the gradient and intercept of the line.

> **TIP**
>
> For part **b**, choose scales so that the points use most of the graph grid (refer to the Skills chapter).

Gradient = ... Intercept = ...

e Write down the equation of your best-fit line using the form $y = mx + c$.

....

f A newton-meter can indicate a force directly, so why is a spring more suitable for applying the horizontal force?

....

....

> **TIP**
>
> For part **d**, the intercept is the value of $\tan \theta$ when L is zero. If your L axis does not start at zero you will have to calculate the intercept value.

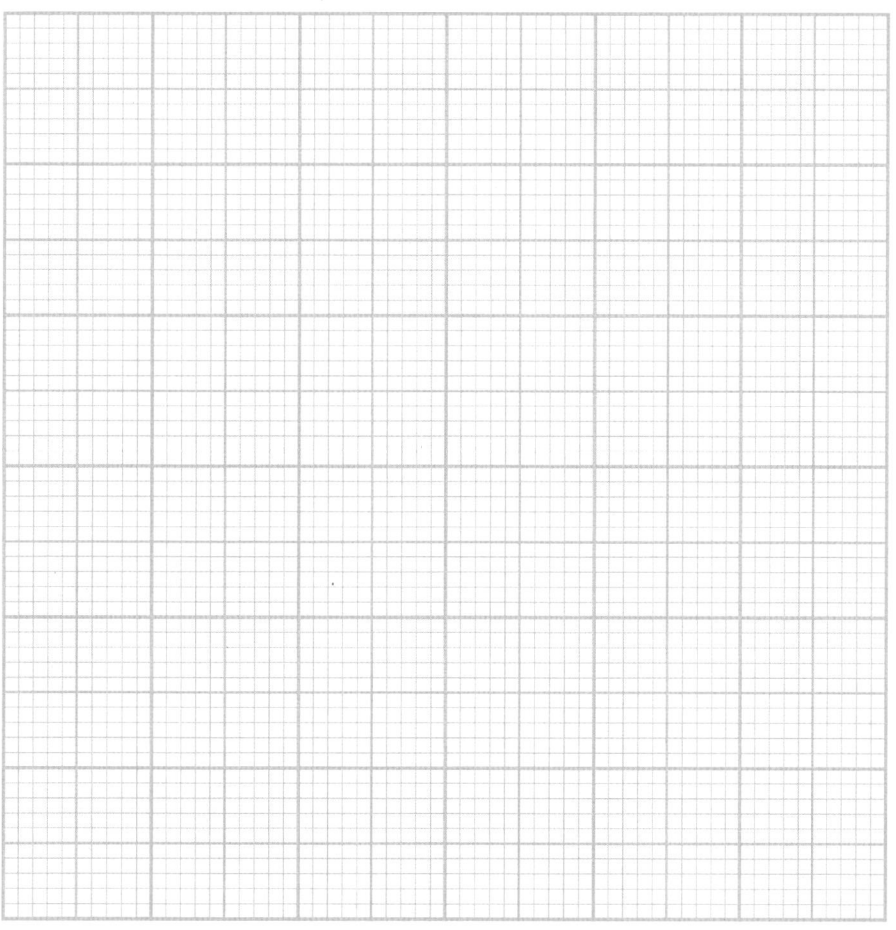

Practical investigation 4.4: Using kinetic energy to do work against friction

When building on soft ground, engineers can drive piles into the ground and then build on top of them.

Figure 4.5 shows a pile driver hammering piles into the ground.

Friction between the pile and the ground provides the force to support the weight of the building.

In this investigation you will test how the friction force varies with the length of pile below ground level.

Figure 4.5: Photograph of a pile driver.

YOU WILL NEED

Equipment:

• tall plastic container filled with rice • ballpoint pen with 1 cm markings along its barrel • 50 g mass hanger with 50 g slotted mass • 30 cm ruler.

Safety considerations

- Make sure you have read the Safety advice at the beginning of this book and listen to any advice from your teacher before carrying out this investigation.

- There are no special safety issues in this investigation.

Method

1   Push the pointed end of the ballpoint pen vertically into the centre of the rice until the 5 cm mark is level with the rice surface.

2   Hold the mass hanger so that it is resting centrally on top of the pen as shown in Figure 4.6.

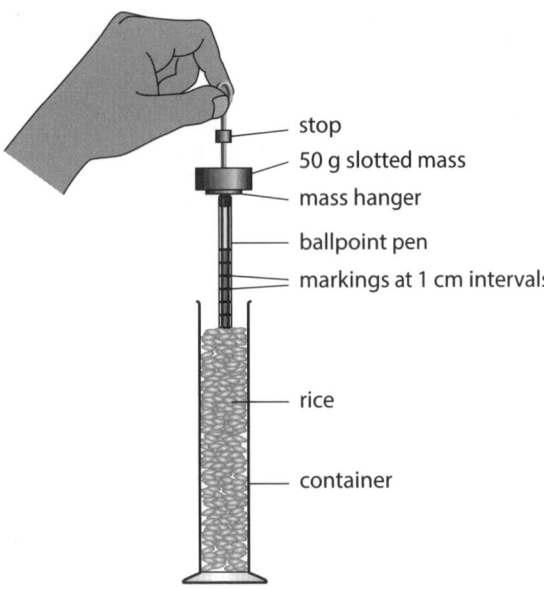

Figure 4.6: Marked ballpoint pen and weight held over a container of rice.

3   Keeping the mass hanger resting on the pen, slide the 50 g mass up to the stop then drop it.

The impact should move the pen further into the rice.

4   Continue to raise and drop the 50 g mass until the next mark is level with the rice surface, keeping a count of the number of impacts, n.

Record n in Table 4.3 in the Results section.

5   Repeat steps **4** and **5** to hammer the pen further, recording the number n of extra impacts needed for each extra centimetre.

6   Measure the distance h that the 50 g mass is lifted before each drop.

Record the value in the Results section.

Results

Depth D / cm	Impacts to give a further 1 cm depth, n	Friction force F / N
5.0		
6.0		
7.0		
8.0		
9.0		
10.0		

Table 4.3: Results table.

$h = $ .... m

TIP

Note that this value of h is in metres.

Analysis, conclusion and evaluation

a   Calculate the energy E transferred by the falling mass in a single impact using

$E = mgh$

where $m = 0.050$ kg and $g = 9.81$ m s^{-2}.

$E = $ .... J

b   For each 1 cm change in depth we can use the fact that:

energy supplied = force × distance moved

Therefore $nE = F \times 1$ cm $= F \times 0.01$ m, where F is the resistance force due to friction.

For each 1 cm change in depth calculate the friction force using:

$$F = \frac{nE}{0.01}$$

Record the values in the last column of Table 4.3.

c   Use the graph grid to plot a graph of F (on the vertical axis) against D (on the horizontal axis).

KEY EQUATIONS

change in gravitational potential energy
$E = mgh$

KEY WORDS

gravitational potential energy: the energy a body has due to its position in a gravitational field

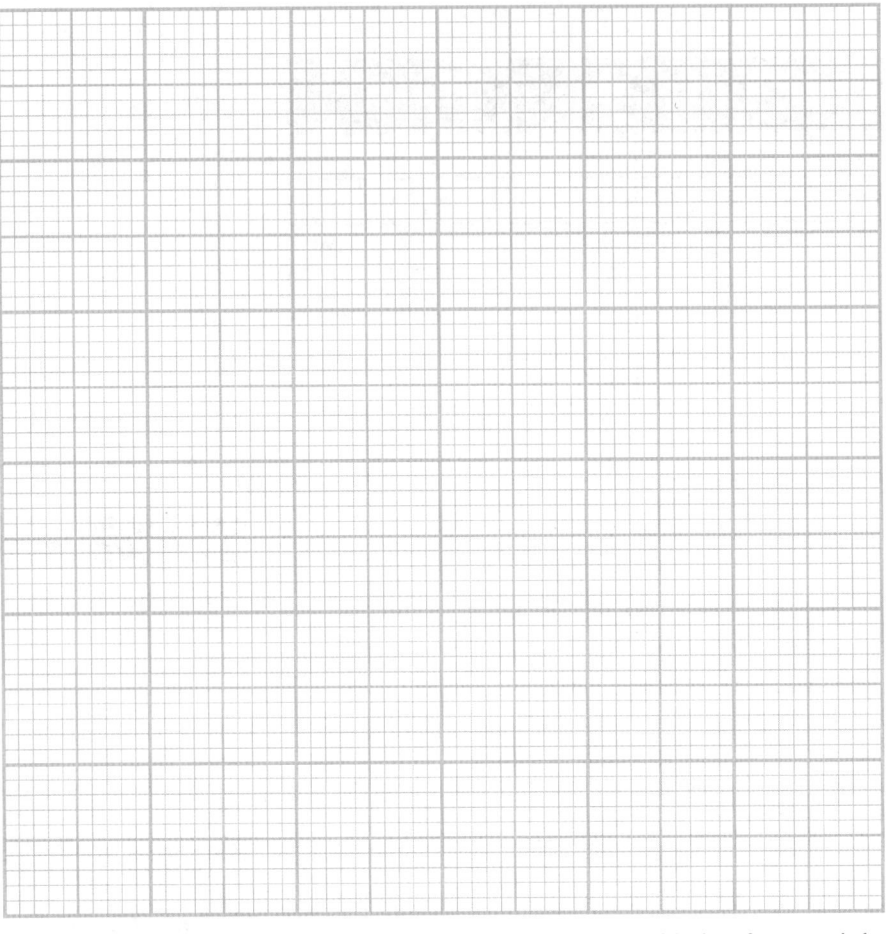

d Describe any relationship that the graph shows between the friction force and the length of the pen below the surface of the rice.

....

....

....

e Describe one feature of the experiment which results in a large uncertainty in the calculated values of F or in the values of D.

....

....

....

Matter and materials

Practical investigation 5.1:
Finding the Young modulus for nylon

The **Young modulus** describes the elastic behaviour of a material. It is used, together with size and shape, to predict how an object will deform under load.

In this practical exercise you will investigate the relationship between the tension force and the extension for a nylon thread, and use the data to find a value for the Young modulus for nylon.

KEY WORDS

Young modulus: the ratio of stress to strain for a given material, provided **Hooke's law** is obeyed

Hooke's law: provided the elastic limit is not exceeded, the extension of an object is proportional to the applied force

YOU WILL NEED

Equipment:
- nylon thread with a loop tied at each end • pulley to fix to edge of table
- wooden block with a hook • G-clamp to clamp the wooden block to the table top • mass hanger • five 100 g slotted masses • metre rule • sticky tape • scissors • micrometer.

Safety considerations

- Make sure you have read the Safety advice at the beginning of this book and listen to any advice from your teacher before carrying out this investigation.

- The nylon thread stores energy as it is stretched, and if it breaks it will spring back very quickly.

- Wear safety goggles when the nylon thread is under tension.

Method

1   Set up the apparatus as shown in Figure 5.1.

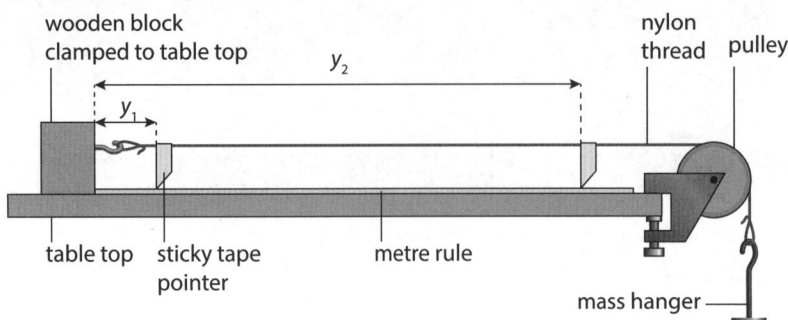

Figure 5.1: Thread over pulley, attached to weight.

2   The two sticky tape pointers should be fixed to the nylon thread approximately 10 cm and 80 cm from the wooden block.

3 y_1 and y_2 are the distances of the two pointers from the wooden block. Take measurements of y_1 and y_2 to the nearest millimetre and record your measurements in the Results section.

4   Add the 100 g masses one at a time to the mass hanger, measuring y_1 and y_2 after each mass has been added. Record your measurements in Table 5.1 in the Results section.

5   Remove the 100 g masses from the mass hanger and then use the micrometer to measure the diameter d of the nylon thread. Take measurements at different places and record your measurements in the Results section. Then calculate and record the average.

Results

For no added mass: $y_1 =$ .... cm $y_2 =$ .... cm

Total added mass M / kg	y_1 / cm	y_2 / cm	Force F /	Extension x /
0.100				
0.200				
0.300				
0.400				
0.500				

Table 5.1: Results table.

TIP

Remember that M does not include the mass of the hanger.

$d_1 =$ .... mm $d_2 =$ .... mm $d_3 =$ .... mm

$d_4 =$ .... mm

Average $d =$ .... mm

Analysis, conclusion and evaluation

a   Using the values of y_2 and y_1 for no added mass, calculate the initial length L using
$$L = y_2 - y_1$$

$L = $ .... cm

b   For each row in Table 5.1, calculate the value of added **force** F using:
$$F = Mg$$
where $g = 9.81$ N m^{-2}.

Add the unit for F to the table heading.

c   For each row in Table 5.1, calculate the value of extension x using:
$$x = y_2 - y_1 - L$$

Add the unit for x to the table heading.

d   Use the graph grid to plot a graph of F (on the vertical axis) against x (on the horizontal axis).

> **KEY EQUATION**
>
> **force** $F = Mg$

> **TIP**
>
> For part **d**, remember to choose simple scales that also ensure a large area of the grid is used.
>
> Remember to label each axis with the variable that is being plotted.

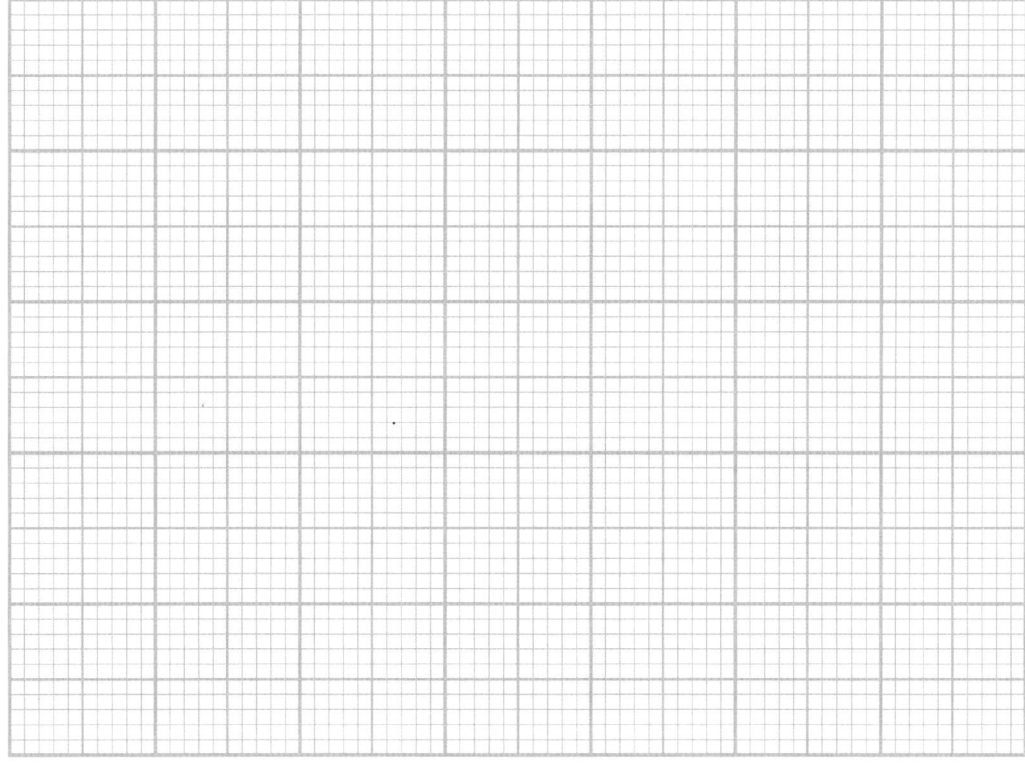

e   Draw the straight line of best fit through the points.

f   Calculate the gradient of the line.

Gradient = ...

g   The gradient is equal to the force constant k for the nylon thread.

Looking carefully at the units of F and x, write down the value of k in N m^{-1}.

k = ... N m^{-1}

h   The Young modulus E for nylon can be calculated using:

$$E = \frac{4LK}{\pi d^2}$$

Calculate E after changing the unit to m for your values for L and d.

L = ... m d = ... m

E = ... N m^{-2}

i   The Young modulus for steel can be found using a similar arrangement of apparatus, but the value of E for steel is much higher than E for nylon.

Explain how a size or quantity could be changed to give values of x similar to those for the nylon thread.

....

....

....

Practical investigation 5.2:
Using a spring to find the Young modulus for steel

A steel coil spring uses the elastic properties of its material. The wire is twisted and bent as the spring is stretched or compressed.

In this practical you will time the oscillations of a mass suspended from a spring, and use your measurements to calculate the Young modulus of the spring material.

Equipment:
• steel spring • digital callipers • 300 g mass with a hook • stand, boss and clamp • stopwatch.

Safety considerations

* Make sure you have read the Safety advice at the beginning of this book and listen to any advice from your teacher before carrying out this investigation.

* There are no special safety issues with this experiment.

Method

1 Use the digital callipers to measure the dimensions of the spring, as shown in Figure 5.2.

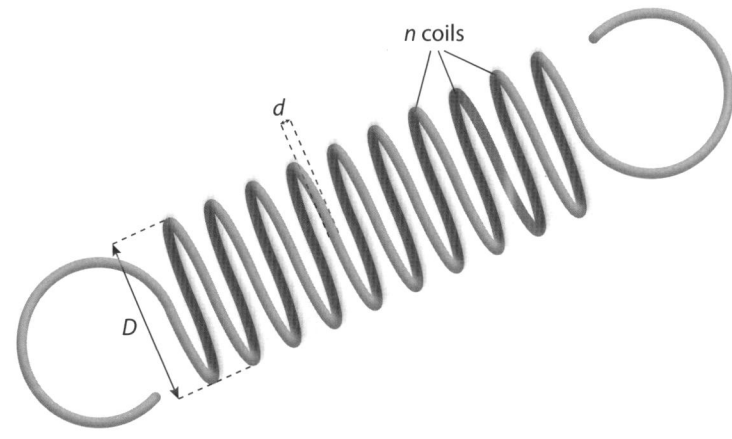

Figure 5.2: Spring.

i   Record the measurements in the Results section, in units of metres, *not* millimetres.

ii Bend the spring to separate the coils and measure the diameter d of the wire. Take the average of several measurements.

iii Measure the outside diameter D of the coiled section. Take the average of several measurements.

iv Count the total number n of coils in the coiled section (do ***not*** include the end loops).

TIP

For step **i**, it is easy to make a mistake in this type of conversion. Divide the mm value by 1000 to get the m value.

2   Hang the spring from the rod of the clamp and then hang the mass from the
     bottom of the spring, as shown in Figure 5.3.

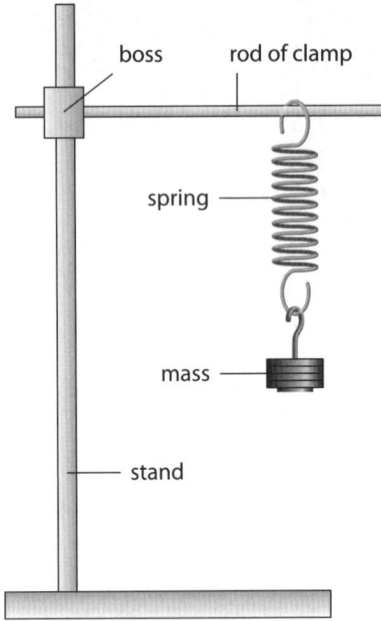

Figure 5.3: Spring with mass attached suspended from rod.

3   Pull the mass down a short distance (about 1 cm). Release it so that it oscillates
     vertically.

     Do this carefully so that the oscillation is vertical, with very little swinging.

4   Measure the time $10T$ for ten *complete* oscillations. Repeat the measurement (to be
     certain that you have not miscounted the oscillations). Record all the $10T$ values
     in Table 5.2.

Results

$d =$ .... m $D =$ .... m $n =$ ....

10T / s			Mean 10T / s	T / s

Table 5.2: Results table.

Analysis, conclusion and evaluation

a   Calculate the mean value of $10T$ and then use it to calculate the value of T.
     Record these values in Table 5.2.

b   The Young modulus E for the steel can be calculated using the formula:

$$E = \frac{490MnD^3}{T^2d^4}$$

where M is the mass suspended from the spring and $M = 0.300$ kg.

Calculate E, recording your value to a suitable number of significant figures.

E = ... N m^{-2}

c   Values of Young modulus are often given using the unit GPa (equal to GN m^{-2}).
    Record your value of E using this unit.

E = ... GPa

d   Using digital callipers means that the diameter d of the wire can be measured
    with a **precision** of 0.01 mm. If Vernier callipers had been used, what would the
    precision have been?

    ....

e   If Vernier callipers had been used, what would the percentage uncertainty have
    been for d?

    ....

    ....

KEY WORD

precision: the smallest change in value that can be measured by an instrument or an operator. A precise measurement is one made several times, giving the same, or very similar, values; there is very little spread about the mean value

Practical investigation 5.3: Water pressure and flow rate

In this practical exercise you will investigate the rate at which water flows out through
a hole in a container, and how it is related to the distance of the hole below the surface
of the water.

YOU WILL NEED

Equipment:
• container with a hole in its base • worktop with sink • jug or beaker of
water • supply of water • stopwatch with lap timer feature • stand, boss and
clamp • metre rule.

Safety considerations

- Make sure you have read the Safety advice at the beginning of this book and listen to any advice from your teacher before carrying out this investigation.

- There are no special safety issues in this investigation.

Method

1   Measure the diameter d at the top of the container.

    Record your average value of d in the Results section.

2   Set up the apparatus as shown in Figure 5.4.

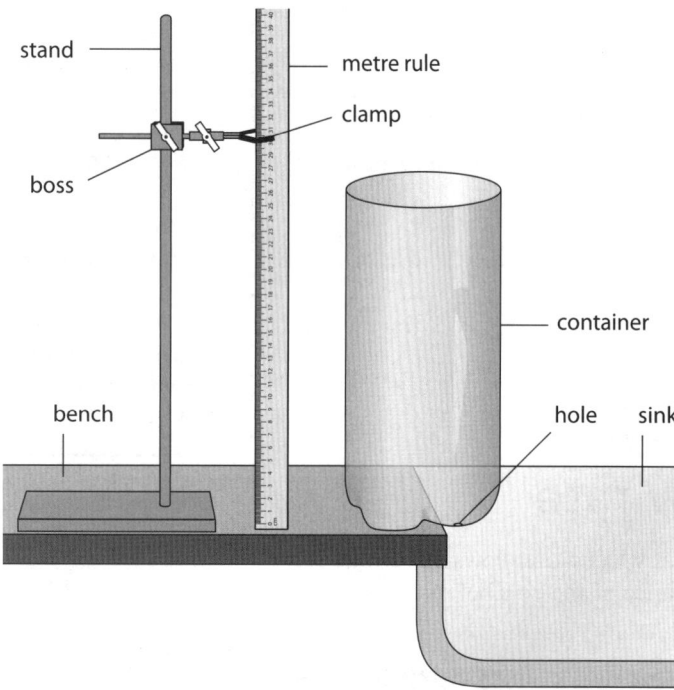

Figure 5.4: Container (with hole) over sink, metre rule.

Ensure the container is positioned so that the hole is above the sink.

The zero end of the metre rule scale should rest on the bench.

3   Pour water from the jug into the container until the level is just below the top, then watch the level fall as water flows out through the hole.

4   Start the stopwatch when the water level passes the 18 cm marking on the metre rule, then practise using the lap timer to take the time at each 2 cm drop in the level (i.e. the time from 18 cm to 16 cm, and the time from 18 cm to 14 cm, and so on).

5 Refill the container until the level is just below the top, then start the stopwatch when the water level passes the 18 cm marking. Measure and record the times for the level to fall to 16 cm, 14 cm, 12 cm, etc.

Record your measurements of water level height h and time T in Table 5.3.

TIP

A timer that can *record* lap times can record all of these values in a single run. Some phone stopwatches have this feature.

Results

Average d = ... cm

Water level height h / cm	Time T / s to fall from 18 cm to h
18.0	0
16.0	
14.0	
12.0	
10.0	
8.0	
6.0	
4.0	

Table 5.3: Results table.

Analysis, conclusion and evaluation

a Calculate the cross-sectional area A of the container using:

$A = \dfrac{\pi d^2}{4}$

A = ... cm^2

b Use the graph grid to plot a graph of h (on the vertical axis) against T (on the horizontal axis).

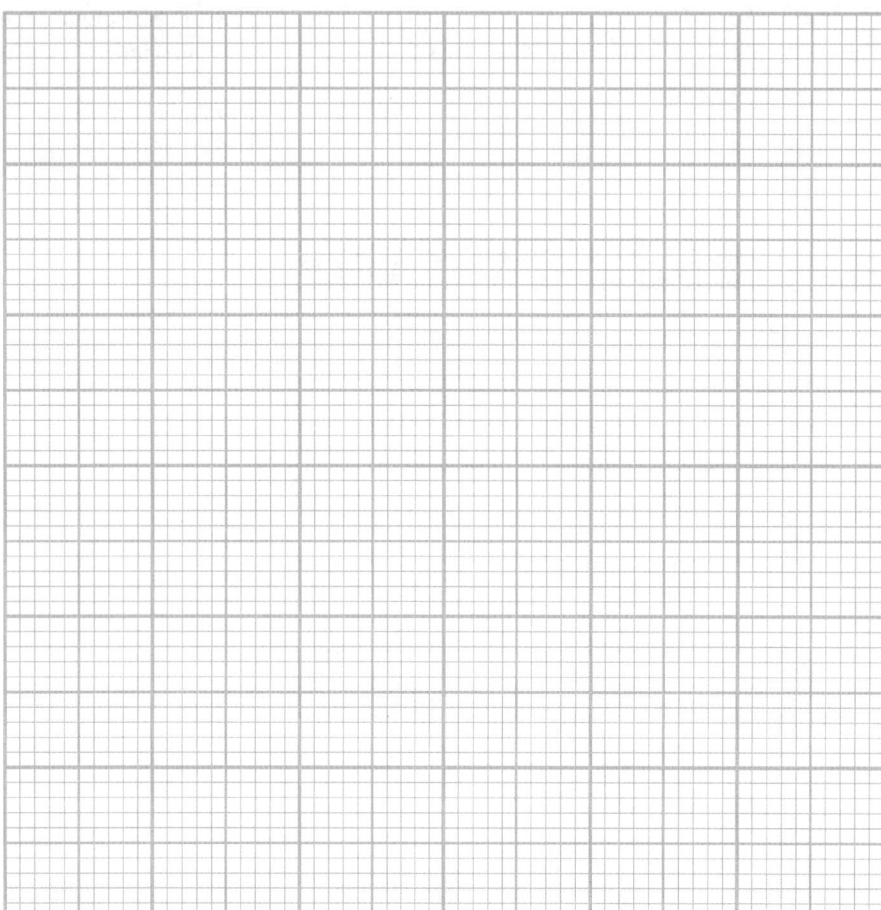

c Draw a smooth curve of best fit through the points.

d Draw the tangent to the curve at $h = 16.0$ cm. The tangent must touch the curve only once, and its line should extend to the edges of the grid.

Calculate the gradient of the tangent. It will have a negative value.

Gradient = ...

e The rate of flow F of water through the hole is given by the cross-sectional area A multiplied by the rate of fall of the water.

Calculate F using:

$F = A \times (-\text{gradient})$

$F =$ .... cm^3 s^{-1}

f Calculate the water pressure P at the hole when $h = 16.0$ cm, using $P = \rho g h$, where $\rho = 1000$ kg m^{-3} and $g = 9.81$ m s^{-2}.

$P = $ .... N m^{-2}

g Repeat parts **d**, **e** and **f** for a value of h of 4.0 cm.

Gradient = ...

$F = $ .... cm^3 s^{-1}

$P = $ .... N m^{-2}

h Test the hypothesis that the water flow rate is proportional to the square root of the water pressure at the hole ($F = k\sqrt{P}$, where k is a constant).

To carry out this test, calculate two values of k, one for each of the two water level heights.

First value of $k = $ ....   Second value of $k = $ ....

Explain whether the two values are close enough to consider the hypothesis valid, including what percentage difference would be acceptable.

....

....

i Describe any difficulty you had in measuring d accurately.

....

....

Describe how a photograph could be used to improve the **accuracy** of the measurement.

....

....

j If the experiment was videoed to make the timing measurements easier, what would have to be in view when the video was replayed?

....

....

> **KEY WORD**
>
> **accuracy:** an accurate measured value of a quantity is close to the true value of the quantity

Chapter 6

Electric current, potential difference and resistance

Practical investigation 6.1: Power and resistance of a lamp

When the voltage across a lamp changes, the current in the lamp changes. The **resistance** of the lamp and the **power** output of the lamp also change.

In this practical investigation you will investigate the relationship between voltage, resistance and power. You will draw graphs of your results.

KEY WORDS

resistance: ratio of the potential difference across the component to the current in the component

power: the rate at which energy is transferred

YOU WILL NEED

Equipment:
- variable power supply • lamp • two digital multimeters
- five connecting leads • switch.

Safety considerations

- Make sure you have read the Safety advice at the beginning of this book and listen to any advice from your teacher before carrying out this investigation.

- Do *not* apply a voltage of more than 6 V across the lamp.

- Switch the circuit off between readings using either the circuit switch or the switch on the power supply.

Method

1   Connect the multimeter across the lamp. Measure the resistance of the lamp. Record the value of resistance in the top row of Table 6.1 in the Results section.

2   Set up the circuit shown in Figure 6.1. The voltmeter should be set in the range 0–20 V.

The ammeter should be set in the range 0–200 mA.

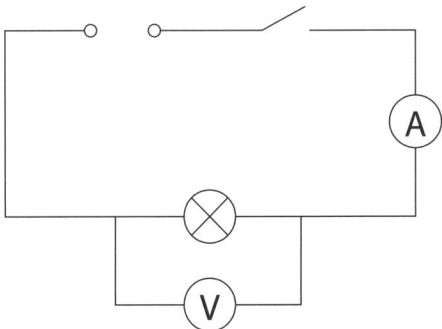

Figure 6.1: Circuit diagram.

3   Close the switch.

Adjust the output of the power supply until the voltmeter reads 1.00 V. Record the ammeter reading I in Table 6.1.

4   Repeat step **3** for the other values of V and record your values of I in Table 6.1.

Results

V / V	I / mA	I /A	R	P
0.00	0.00	0.00	-	0.000
1.00				
2.00				
3.00				
4.00				
5.00				
6.00				

Table 6.1: Results table.

> **TIP**
>
> It is not important if the reading is not exactly 1.00 V as long as a wide range of readings is taken using the apparatus. There is room in Table 6.1 to record your values of V.

> **TIP**
>
> If I is 40 mA, then I is 0.040 A.

Analysis, conclusion and evaluation

a   Calculate values of resistance R and power P using:

$R = \dfrac{V}{I}$ and $P = V \times I$

and record your values in Table 6.1. You need to add the units to the column headings as well.

KEY EQUATIONS

resistance $= \dfrac{\text{potential difference}}{\text{current}}$, $R = \dfrac{V}{I}$

power $=$ potential difference $\times$ current, $P = V \times I$

b   Plot a graph of R on the y-axis against V on the x-axis using the graph grid.

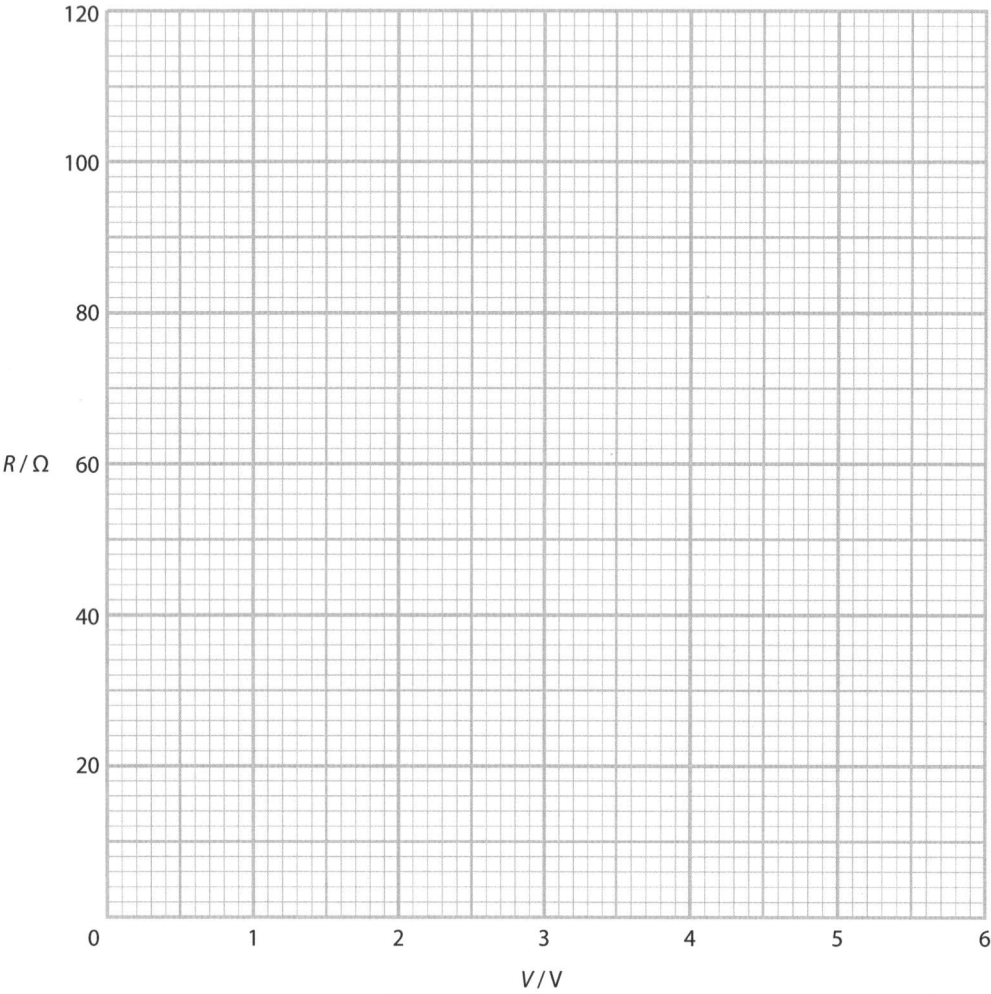

c   Draw the curve of best fit through your points.

d   Plot a graph of P on the y-axis against V on the x-axis using the graph grid.

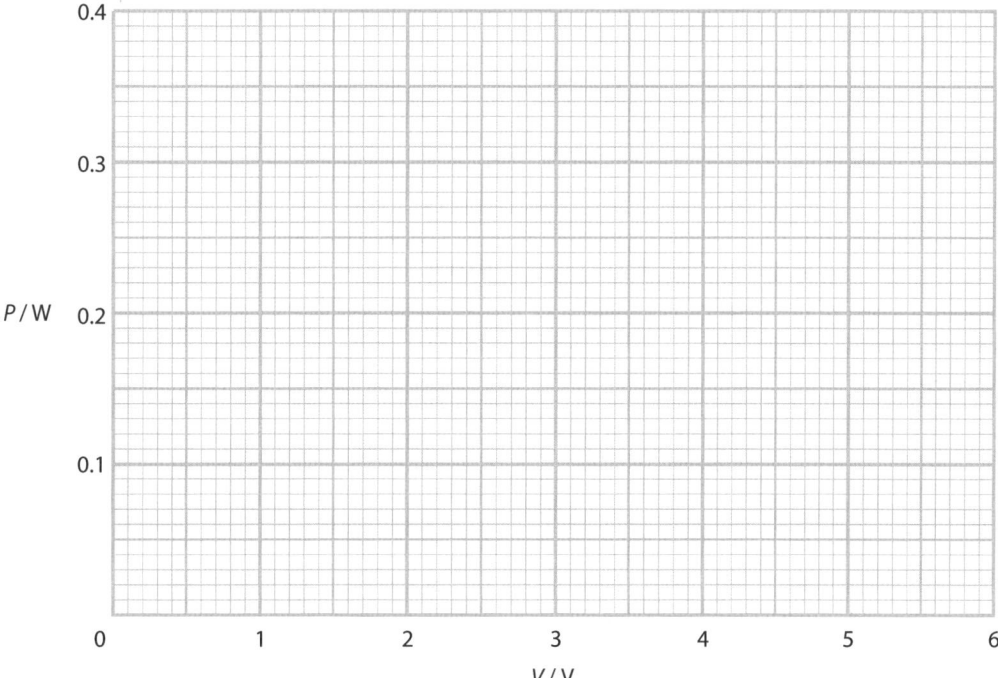

e   Draw the curve of best fit through your points.

f   How does R vary with V?

....

....

g   How does P vary with V?

....

....

h   Would you expect the R–V or the P–V graph to go through the point (0, 0)?
      Give your reasons.

....

....

....

....

i   There are three forms of equation for the power dissipated in a lamp:

$$P = VI, P = \frac{V^2}{R} \text{ and } P = I^2 R$$

KEY EQUATIONS

power $P = VI, P = \dfrac{V^2}{R}$ and $P = I^2 R$

The lamp you have used might be labelled '6 V 60 mA' or '6 V 0.36 W'.

Complete the two sketch graphs in Figure 6.2 to suggest how R varies with V and P varies with V for a lamp that is labelled '240 V 100 W' for the range 0–240 V.

Add suitable scales to the axes.

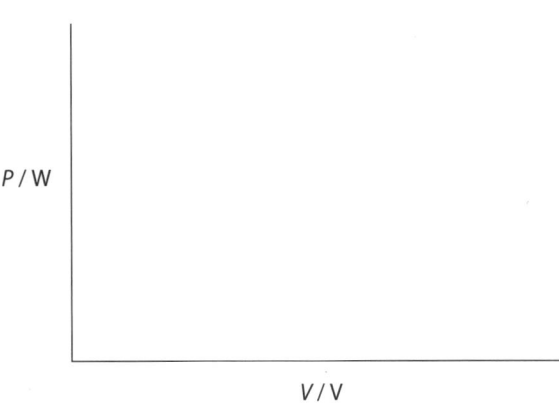

Figure 6.2: Two sets of axes.

Practical investigation 6.2: Resistors in series

When a resistor is connected to another resistor in series the total resistance in the circuit changes, so the current in the circuit also changes.

In this practical exercise you will investigate the relationship between the added resistors and the current.

You will draw a graph of your results and use both intercepts to determine the values of some of the circuit components.

YOU WILL NEED

Equipment:

- 1.5 V cell • switch • five connecting leads • two component holders
- digital multimeter • resistors with the following values: 18 Ω, 22 Ω, 27 Ω, 33 Ω and 2 × 15 Ω.

Safety considerations

- Make sure you have read the Safety advice at the beginning of this book and listen to any advice from your teacher before carrying out this investigation.

- Open the switch between readings. If the switch remains closed for a long period of time the **electromotive force (e.m.f.)** of the cell will drop.

KEY WORDS

electromotive force (e.m.f.): the amount of energy changed from other forms into electrical energy per unit charge produced by an electrical supply

Method

1   Set up the circuit shown in Figure 6.3. Both resistors X and R should have values of resistance equal to 15 Ω. The multimeter should be set in the range 0–200 mA.

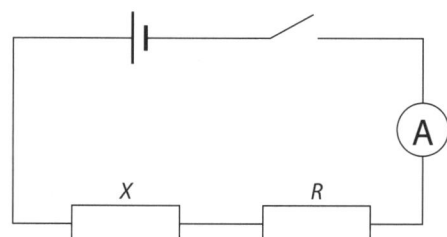

Figure 6.3: Circuit diagram.

2   Close the switch. Record the ammeter reading I in Table 6.2 in the Results section.

3   Remove the 15 Ω resistor from R and replace it with the 18 Ω resistor.

Do not change the resistance X which should have a value of 15 Ω throughout the experiment.

4   Close the switch. Record the ammeter reading in Table 6.2.

5   Repeat step **2** using the other resistors and record your values of R and I in Table 6.2.

6   Disconnect the circuit. Connect the voltmeter to the cell. Record the value of its e.m.f. E in the Results section.

Results

$E =$ V

$R\ /\ \Omega$	$I\ /\ mA$	$I\ /\ A$	$\frac{1}{I}\ /\ A^{-1}$
15			
18			
22			
27			
33			

Table 6.2: Results table.

Analysis, conclusion and evaluation

a How does I vary with R?

....

b Calculate values of $\frac{1}{I}$ and write them in Table 6.2. You need to add the units to the column heading as well.

c Plot a graph of $\frac{1}{I}$ on the y-axis against R on the x-axis using the graph grid.

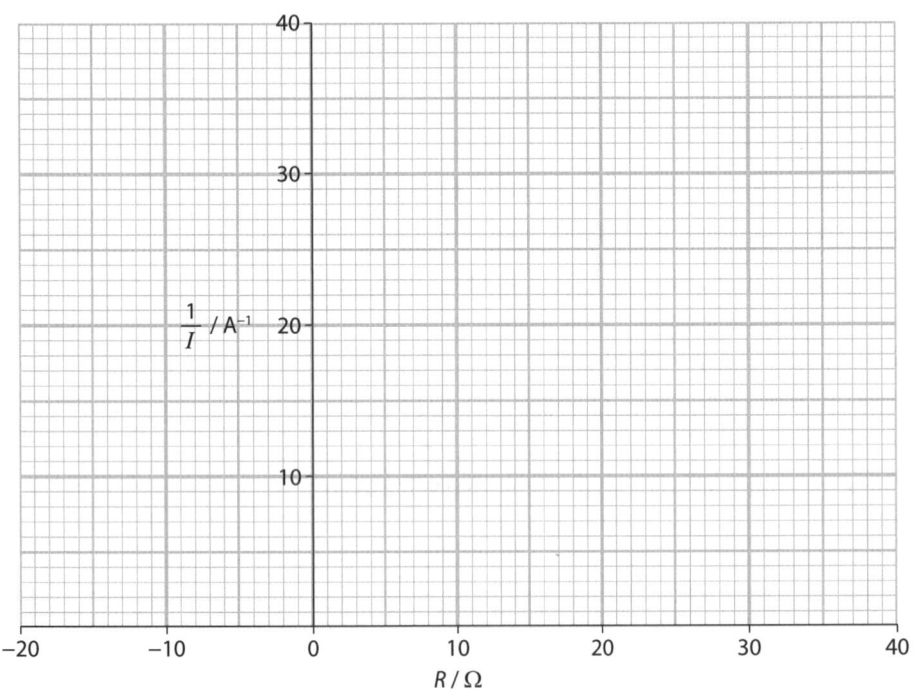

TIP

If I is 150 mA, then I is 0.150 A and $\frac{1}{I}$ is 6.67 A⁻¹.

d Draw the straight line of best fit through your points.

e Determine the gradient of the line.

Gradient = ...

f The total resistance in the circuit is equal to $R + X$.

The total resistance is also equal to $\dfrac{E}{I}$, where E is the e.m.f. of the cell.

So $\dfrac{E}{I} = R + X$

Therefore $\dfrac{1}{I} = \dfrac{R}{E} + \dfrac{X}{E}$

A graph of $\dfrac{1}{I}$ against R will have gradient $= \dfrac{1}{E}$

Use the gradient of your graph to determine E.

$E =$ V

g The y-intercept of the graph of $\dfrac{1}{I}$ against R is $\dfrac{X}{E}$.

Determine the y-intercept of your graph.

y-intercept = ...

h Use the y-intercept to determine X.

$X =$ ....

i The x-intercept of the graph of $\dfrac{1}{I}$ against R will be when

$\dfrac{1}{I} = 0$ or $\dfrac{R}{E} = \dfrac{X}{E}$ or $R = -X$

Determine the x-intercept of your graph.

x-intercept = ...

j Use the x-intercept to determine X.

$X =$ ....

> **TIP**
>
> Note that it is impossible for $\dfrac{1}{I}$ to be zero because I would have to be infinitely large. It is also impossible to have negative values of R.

k   Suppose the experiment was repeated using the same cell but with $X = 10\ \Omega$. How would this affect the values of the gradient and the intercepts? Mark your answers in Table 6.3 with a tick (✓).

	Smaller	Same	Bigger
Gradient			
y-intercept			
x-intercept			

Table 6.3: Effects of changing X.

l   Draw the line for $X = 10\ \Omega$ on your grid and label it 'Y'.

m   In step **6** you measured the e.m.f. of the cell using the voltmeter. Calculate the percentage difference between the measured value of E and the value of E obtained from the graph.

....

....

n   Suggest a reason why the value of X found from the x-intercept might be unreliable.

....

....

Practical investigation 6.3: Resistors in parallel

When a resistor is connected to another resistor in parallel the total resistance in the circuit changes. The current in the circuit changes.

In this practical exercise you will investigate the relationship between the added resistor and the current.

You will draw a graph of your results and use both intercepts to determine the values of some of the circuit components.

> **YOU WILL NEED**
>
> **Equipment:**
> • 1.5 V cell • switch • six connecting leads • two component holders
> • digital multimeter • resistors with the following values: 120 Ω, 150 Ω,
> 180 Ω, 220 Ω and 2 × 100 Ω.

Safety considerations

- Make sure you have read the Safety advice at the beginning of this book and listen to any advice from your teacher before carrying out this investigation.

- Switch off between readings using the circuit switch.

Method

1   Set up the circuit shown in Figure 6.4. Both resistors *X* and *R* should have values of resistance equal to 100 Ω. The multimeter should be set in the range 0–200 mA.

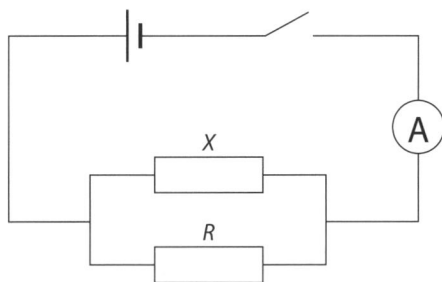

Figure 6.4: Circuit diagram.

2   Close the switch. Record the ammeter reading *I* in Table 6.4 in the Results section.

3   Remove the 100 Ω resistor from *R* and replace it with the 120 Ω resistor.

Do not change the resistance *X*, which should have a value of 100 Ω throughout the experiment.

Close the switch and record the ammeter reading in Table 6.4.

4   Repeat step **3** using the other resistors and write your values for *R* and *I* in Table 6.4.

5   Disconnect the circuit. Connect the voltmeter to the cell. Record the value of its e.m.f. *E* in the Results section.

> **TIP**
>
> If *I* is 25 mA, then *I* is 0.025 A.

Results

$E =$ V

R / Ω	I / mA	I / A	
100			
120			
150			
180			
220			

Table 6.4: Results table.

> ### TIP
>
> The values of resistance are greater in this experiment than in the previous experiment. There are two reasons:
>
> - When the resistors are connected in parallel, the total resistance decreases so the current increases and the cell is discharged at a greater rate.
>
> - When the total resistance in a circuit is low, the resistance of the connecting leads becomes significant.

Analysis, conclusion and evaluation

a   How does I vary with R?

....

....

b   Calculate values of IR in volts and write them in Table 6.4.

c   Plot a graph of IR on the y-axis against R on the x-axis using the graph grid.

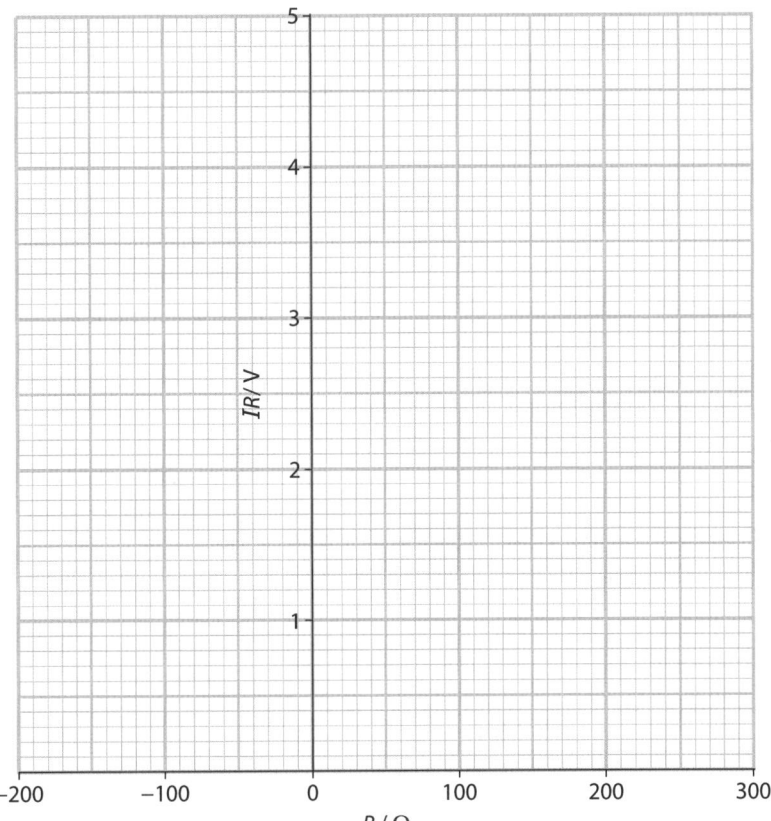

d   Draw the straight line of best fit through your points.

e   Determine the gradient of the line.

Gradient = ...

f   The total resistance in the circuit is equal to $\dfrac{RX}{R+X}$

The total resistance is also equal to $\dfrac{E}{I}$, where E is the e.m.f. of the cell.

So $\dfrac{E}{I} = \dfrac{RX}{R+X}$

Therefore $IR = \dfrac{ER}{X} + E$

A graph of IR against R will have gradient $= \dfrac{E}{X}$ and y-intercept $= E$.

Determine the y-intercept of your graph.

y-intercept = ...

Use the y-intercept of your graph to determine E.

$E =$ V

g Use the gradient of your graph to determine X.

$X =$ .... Ω

h The x-intercept will be the value of R when $IR = 0$ or $\dfrac{ER}{X} = -E$ or $R = -X$

Use the x-intercept of your graph to find another value for X.

$X =$ .... Ω

i Suppose the experiment was repeated using the same cell but with $X = 50\ \Omega$. What would be the effect on the gradient and the intercepts? Mark your answers in Table 6.5 with a tick (✓).

	Smaller	Same	Bigger
Gradient			
y-intercept			
x-intercept			

Table 6.5: Effects of changing X.

j Draw the line for $X = 50\ \Omega$ on your grid and label it 'Z'.

k In step **5** you measured the e.m.f. of the cell using the voltmeter. Calculate the percentage difference between the measured value of E and the value of E obtained from the graph.

....

....

l Suggest one reason why the value of X found from the x-intercept might be unreliable.

....

....

> **TIP**
>
> Note that it is impossible for IR to be zero because I or R would have to be zero. It is also impossible to have negative values of R.

Chapter 7

Resistance and resistivity

CHAPTER OUTLINE

This chapter relates to Chapter 10: Resistance and resistivity and Chapter 11: Practical circuits, in the coursebook.

In this chapter you will complete investigations on:

- 7.1 Resistivity of the metal of a wire

- 7.2 Internal resistance of a dry cell

- 7.3 Potential divider.

Practical investigation 7.1:
Resistivity of the metal of a wire

The **resistance** R of a metal wire depends on its length l, cross-sectional area A and the **resistivity** ρ of the metal from which it is made.

In this practical exercise you will investigate the relationship between resistance and length.

You will draw a graph of your results. You will measure the diameter of the wire and use your results to determine the resistivity of the metal.

KEY EQUATION

$$\text{resistance} = \frac{\rho \times l}{A}$$

YOU WILL NEED

Equipment:
- two connecting leads • two crocodile clips • digital multimeter • metre rule.

Access to:
- micrometer • reel of resistance wire • adhesive tape • scissors • wire cutters.

KEY WORD

resistivity: a measure of electrical resistance, defined as resistance x cross-sectional area / length

Safety considerations

- Make sure you have read the Safety advice at the beginning of this book and listen to any advice from your teacher before carrying out this investigation.

- There are no other specific safety issues with this investigation.

Method

1   Use the wire cutters to cut a wire of length 110 cm.

2   Use the scissors to cut sufficient tape to attach the wire to the metre rule as shown in Figure 7.1.

Figure 7.1: Wire attached with tape to ends on a metre rule.

3   Set up the circuit shown in Figure 7.2. The **ohmmeter** should be set to an appropriate range. The distance l between the crocodile clips should be 0.100 m. Record the value of the ohmmeter reading in Table 7.1 in the Results section.

> **KEY WORD**
>
> **ohm:** the unit of electric resistance measured by an ohmmeter

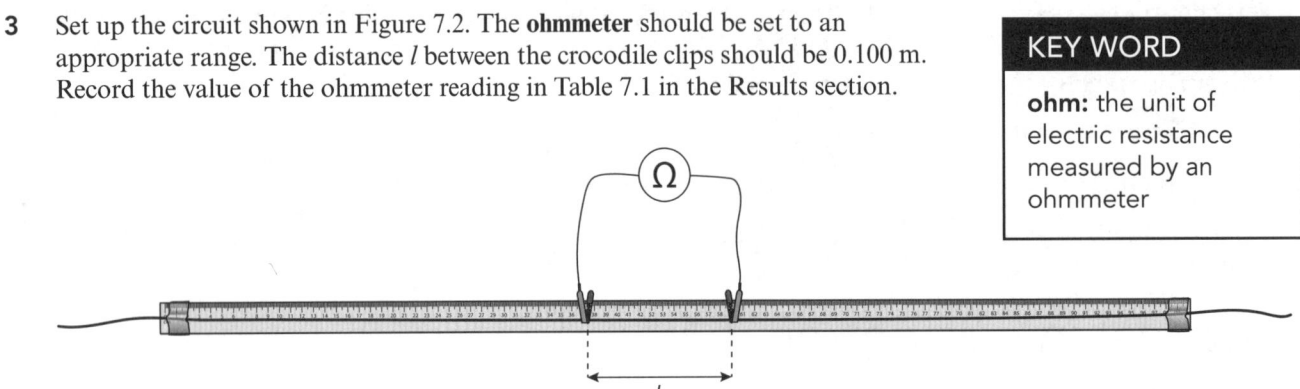

Figure 7.2: As Figure 7.1, but with ohmmeter.

4   Repeat step **3** for the other values of l and record your values of R in Table 7.1.

5   Measure the diameter d of the wire using the micrometer. Record your value in the Results section.

6   Connect the connecting leads in series and use the ohmmeter to measure the total resistance of these leads. Record the value of this resistance in the Results section.

7   Do not remove the wire from the metre rule. You will need to use this again in Practical investigation 7.3.

Results

l / m	R / Ω
0.100	
0.250	
0.400	
0.550	
0.700	
0.850	

Table 7.1: Results table.

Diameter d = ... mm

Total resistance of the connecting leads = ... Ω

Analysis, conclusion and evaluation

a Calculate the cross-sectional area A of the wire using:

$$A = \frac{\pi d^2}{4}$$

TIP

Measure and record d in mm but, to calculate A, convert d to m because values of resistivity are quoted with the unit ohm metre (Ω m).

$A =$ .... m^2

b Plot a graph of R on the y-axis against l on the x-axis using the graph grid.

TIP

Note that $1\ mm^2 = 10^{-6}\ m^2$.

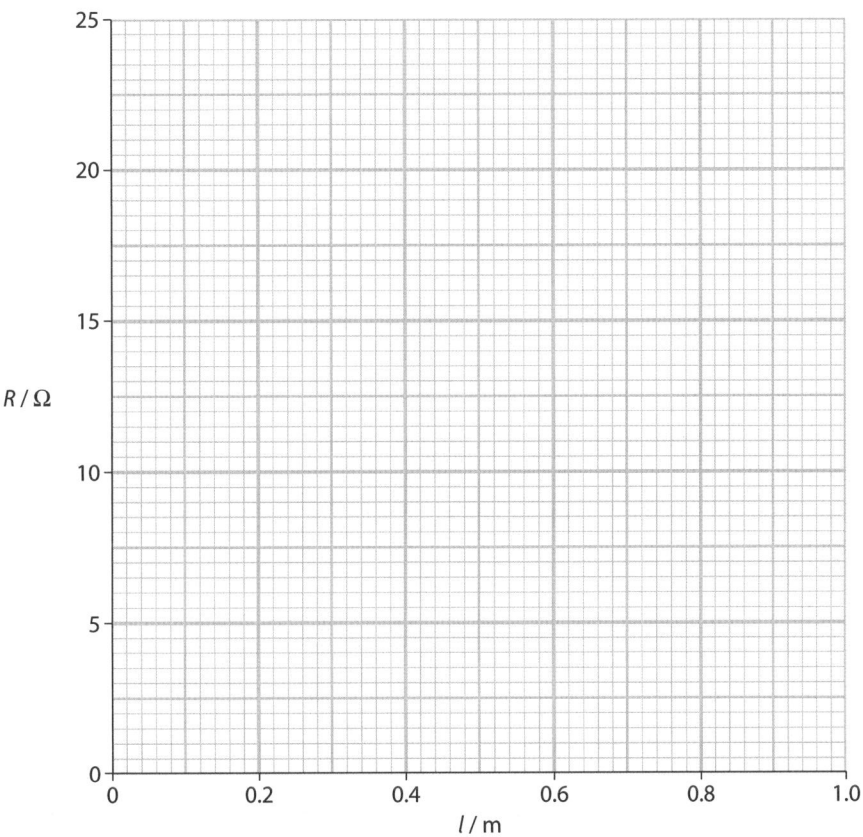

c Draw the straight line of best fit through your points.

d Determine the gradient of your line.

Gradient = ...

e Determine the y-intercept of your line.

y-intercept = ...

f The relationship between R, ρ, l and A is

$$R = \frac{\rho l}{A}$$

The gradient of your graph is $\dfrac{R}{l}$

So $\rho = \text{gradient} \times A$

Determine a value for ρ.

ρ = ... $\Omega\,m$

g Wires used in schools are usually made from constantan, nichrome or copper. These metallic materials have the following values of resistivity and diameter.

$\rho_{constantan}$	$4.9 \times 10^{-7}\ \Omega\,m$
$\rho_{nichrome}$	$1.2 \times 10^{-6}\ \Omega\,m$
ρ_{copper}	$1.7 \times 10^{-8}\ \Omega\,m$
diameter / mm	0.46, 0.38, 0.32, 0.27, 0.23, 0.19, 0.15

Circle the material and the diameter that you have used in this experiment.

h If you used constantan wire, calculate the diameter of nichrome wire that would have given you similar results.

or If you used nichrome wire, calculate the diameter of constantan wire that would have given you similar results.

Diameter = ... mm

i Show why copper wire would be unsuitable for this experiment.

....

....

j A graph that is a straight line that does ***not*** go through the point (0, 0) and has a positive intercept on the y-axis suggests that there is a systematic error in the readings.

This could be due to the resistance of the connecting leads.

Use your results to investigate this idea.

....

....

....

Practical investigation 7.2: Internal resistance of a dry cell

When a voltmeter is connected across a dry cell the reading on the voltmeter is the e.m.f. of the cell. Little or no current is drawn from the cell because the voltmeter has a very high resistance.

When the cell is connected in a circuit where current is drawn from the cell the voltmeter reading drops. This is because some of the volts are 'lost' across the internal resistance of the cell.

In this practical exercise you will investigate the relationship between the **potential difference (p.d.)** across the cell and the current drawn from the cell.

You will draw a graph of your results and use the gradient to determine the internal resistance of the cell.

Finally, you will investigate whether the resistances of the connecting leads are significant.

KEY WORDS

potential difference (p.d.): the potential difference, V, between two points, A and B, is the energy transferred per unit charge as it moves from point A to point B

YOU WILL NEED

Equipment:
- 1.5 V cell • switch • six connecting leads • two digital multimeters
- rheostat.

Safety considerations
- Make sure you have read the Safety advice at the beginning of this book and listen to any advice from your teacher before carrying out this investigation.

- There are no particular safety issues with this investigation.

Method
1   Connect the voltmeter across the dry cell to measure the e.m.f. E of the cell. Record the reading in the Results section.

2   Set up the circuit shown in Figure 7.3. The switch should be open. Adjust the scale on the voltmeter to read 0–2 V to the nearest 0.001 V. Adjust the scale on the ammeter to read 0–200 mA to the nearest 0.1 mA.

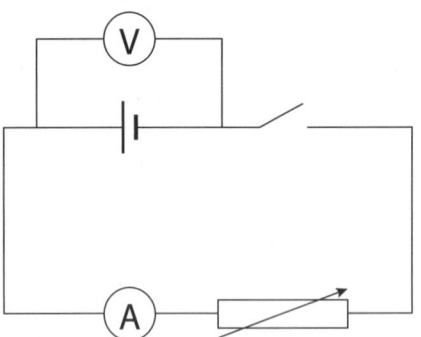

Figure 7.3: Circuit diagram.

3 Close the switch and adjust the position of the slider on the rheostat until the ammeter reading is a maximum but below 200 mA. Open the switch.

4 Read the voltmeter and record its reading in the top row of Table 7.2 in the Results section.

5 Close the switch and record the values of V and I in Table 7.2.

6 Open the switch and wait until the voltmeter reaches its maximum value. Record this reading in Table 7.2.

7 Adjust the rheostat so that less current is drawn from the cell. Repeat steps **5** and **6**.

8 Disconnect the circuit and measure the resistance of each of the connecting leads. Record your results in Table 7.3 in the Results section.

9 Repeat step **1** and wait at least 30 minutes before reading the voltmeter. Record this final voltmeter reading in the Results section.

Results

E = ... V

Maximum voltmeter reading with switch open / V	I / mA	I / A	V / V

Table 7.2: Results table.

Final voltmeter reading = ... V

Resistance of connecting leads / Ω					

Table 7.3: Results table.

Analysis, conclusion and evaluation

a   How does *V* vary with *I*?

....

b   Plot a graph of *V* on the *y*-axis against *I* on the *x*-axis using the graph grid.

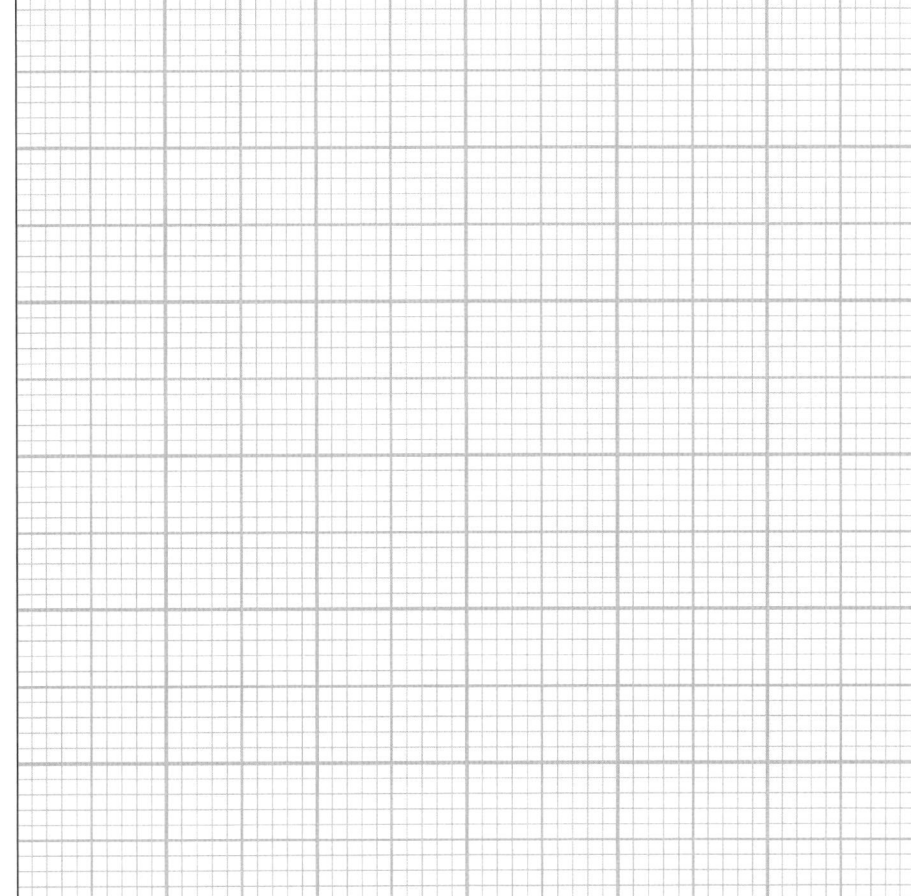

I / A

V / V

c   Draw the straight line of best fit through your points.

d   Determine the gradient of the line.

Gradient = ...

e Using **Kirchhoff's second law**, the e.m.f. of the cell is equal to the sum of the potential differences across all the resistors in the circuit.

So $E = V + Ir$

where r is the internal resistance of the cell.

Therefore $V = -Ir + E$

A graph of V against I will have gradient $= -r$

Use the gradient of your graph to determine r.

$r = \dots\dots\dots\dots\dots \Omega$

f The y-intercept of the graph of V against I is E.

Determine the y-intercept of your graph.

y-intercept $= \dots\dots\dots\dots\dots$

g Use the y-intercept to determine E.

$E = \dots\dots\dots\dots\dots V$

h In steps **1** and **9** you measured the e.m.f. of the cell using the voltmeter. Comment on the difference between the measured values of E and the value of E obtained from the graph.

....

....

i In step **8** you measured the resistance of each of the connecting leads. Comment on whether these resistances are significant in this experiment.

....

....

....

....

....

j You are often asked to open the switch in a circuit between readings. Why is this advisable?

....

....

....

....

> **KEY WORDS**

Kirchhoff's second law: the sum of the e.m.f.s around a closed loop is equal to the sum of the p.d.s in that same loop; this law represents the conservation of energy

> **KEY EQUATION**

Kirchhoff's second law
$E = V + Ir$

Practical investigation 7.3: Potential divider

A **potential divider** circuit allows you to use part of the e.m.f. of a power supply as an output.

In this practical exercise you will connect a resistor and a wire in series across a power supply and investigate the relationship between the output voltage across the resistor and the length of the wire.

You will draw a graph of your results and use the gradient and the y-intercept to determine the value of the e.m.f. of the power supply and the resistance of the resistor.

KEY WORDS

potential divider: a circuit in which two or more components are connected in series to a supply; the output voltage is taken across one of the components

YOU WILL NEED

Equipment:

- the wire on the metre rule used in Practical investigation 7.1 • 1.5 V cell
- switch • six connecting leads • two crocodile clips • 15 Ω resistor
- component holder for the resistor • digital multimeter to measure 0–2 V to the nearest 0.001 V.

Safety considerations

- Make sure you have read the Safety advice at the beginning of this book and listen to any advice from your teacher before carrying out this investigation.

- Switch off between readings using the circuit switch.

- There are no particular safety issues with this investigation.

Method

1   Set the multimeter in the range 0–2 V and measure the e.m.f. E of the cell. Record your initial value in the Results section.

2   Set up the circuit shown in Figure 7.4. The multimeter should be set in the range 0–2 V. The distance l between the crocodile clips attached to the wire should be 0.200 m.

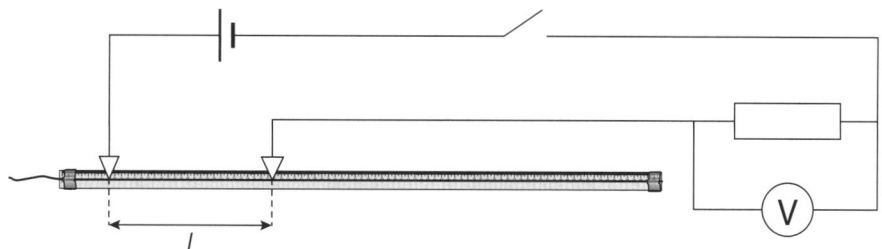

Figure 7.4: Circuit diagram with wire on metre rule.

3 Close the switch. Record the voltmeter reading V in Table 7.4. Open the switch.

4 Repeat step **3** using the other values of l and write your values of V in Table 7.4.

5 Disconnect the circuit. Connect the voltmeter to the cell. Record its e.m.f. E. Record this final value in the Results section.

Results

Initial value of E = ... V

l / m	V / V	$\dfrac{1}{V}$
0.200		
0.300		
0.400		
0.500		
0.600		
0.700		
0.800		

Table 7.4: Results table.

Final value of E = ... V

Analysis, conclusion and evaluation

a How does V vary with l?

....

....

b Calculate values of $\dfrac{1}{V}$ and write them in Table 7.4. You need to add the units to the column heading as well.

c Plot a graph of $\dfrac{1}{V}$ on the y-axis against l on the x-axis using the graph grid on the next page.

d Draw the straight line of best fit through your points.

e Determine the gradient of the line.

Gradient = ...

f The total resistance in the circuit is equal to $\frac{\rho l}{A} + R$, neglecting the internal resistance of the cell.

Therefore $\frac{1}{V} = \frac{\rho l}{ARE} + \frac{1}{E}$

A graph of $\frac{1}{V}$ against l will have gradient $\frac{\rho}{ARE}$ and y-intercept $\frac{1}{E}$.

Determine the y-intercept of your graph.

y-intercept = ...

g Use the y-intercept to determine E.

$E = $ V

h Use the y-intercept and gradient of your graph to determine R.

$R = $ .... Ω

i You have recorded three values for E.
Explain any differences between these values.

....

....

....

....

....

....

....

> **KEY EQUATION**
>
> **Using the potential divider equation:**
>
> $$V = \frac{ER}{\frac{\rho l}{A} + R}$$

> Chapter 8

Waves

CHAPTER OUTLINE

This chapter relates to Chapter 12: Waves and Chapter 14: Stationary waves, in the coursebook.
In this chapter you will complete investigations on:

* 8.1 Stationary waves on a wire carrying a current

* 8.2 Inverse-square law for waves from a point source

* 8.3 Refraction of light waves by a lens.

Practical investigation 8.1: Stationary waves on a wire carrying a current

The formation of **stationary waves** on a string is discussed in the Cambridge International AS & A Level Physics coursebook. In this experiment, the wire is placed in a magnetic field and made to vibrate by passing an alternating current through it so that the **frequency** of the vibration is the same as the frequency of the a.c. supply.

The experiment investigates the relationship between the tension in the wire and the **wavelength** of the stationary wave.

YOU WILL NEED

Equipment:
* pulley wheel • wire with one end clamped to the bench • prism
* steel yoke with two ceramic magnets attached to it • 2 V a.c. power supply
* two connecting leads, each with a clip at one end • 100 g mass hanger
* two 100 g slotted masses and a 50 g slotted mass • metre rule
* sheet of dark-coloured paper.

Safety considerations

* Make sure you have read the Safety advice at the beginning of this book and listen to any advice from your teacher before carrying out this investigation.

* Wear safety goggles when the wire is under tension.

* There is no safety issue with the very low voltage electrical supply.

KEY WORDS

stationary wave: a wave pattern produced when two progressive waves of the same frequency travelling in opposite directions combine

frequency: the number of oscillations per unit time

wavelength: the distance between two adjacent peaks or troughs in a wave or the distance between adjacent points having the same phase

Method

1   Set up the apparatus as shown in Figure 8.1.

2   The wire and pulley have been set up for you with a 100 g mass hanger at the end of the wire.

3   Position the steel yoke and magnets approximately 15 cm from the wooden blocks and with the wire between the magnets.

4   Position the prism under the wire. The prism should be in contact with the wire.

5   Use clips and leads to connect the a.c. power supply to the wire at the points shown in Figure 8.1.

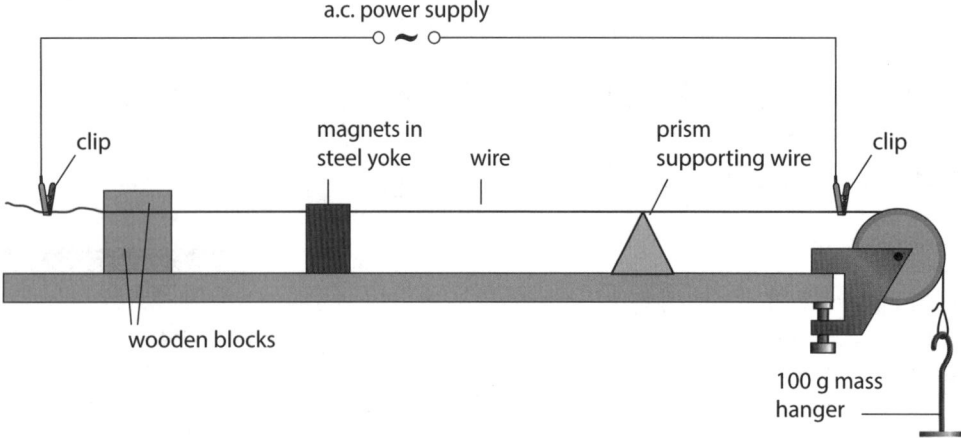

Figure 8.1: Circuit diagram with wire over pulley.

6   *M* is the mass hanging from the wire. Record the value of *M* in Table 8.1 in the Results section.

7   Switch on the power supply.

8   Slowly move the prism along under the wire until the wire vibrates as shown in Figure 8.2. Adjust the prism position until the amplitude of the vibration is as large as possible.

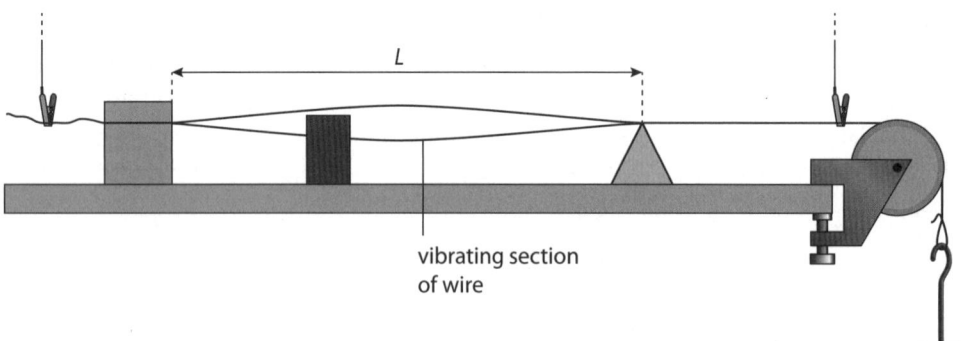

Figure 8.2: Wire over pulley vibrating.

9 *L* is the length of wire between the prism and the wooden blocks as shown in Figure 8.2. Measure *L* and record the value in Table 8.1.

10 Switch off the power supply.

11 Increase *M* in steps of 50 g, repeating steps **3** to **10** after each increase.

TIP

The vibration will be easier to see if the black paper is held behind the wire.

Results

M / kg	L / m	λ / m	λ² / m²

Table 8.1: Results table.

TIP

Although *L* is recorded in metres, it must be measured to the nearest mm (0.001 m).

Analysis, conclusion and evaluation

a   The wavelength of the stationary wave is λ and this is twice the distance between adjacent **nodes**. Calculate and record λ for each row in Table 8.1 using the relationship $\lambda = 2L$

KEY WORD

node: a point on a stationary wave where the amplitude is zero

b   For each row in Table 8.1 calculate and record the value of λ^2.

Ensure that every column in Table 8.1 has a heading with a quantity and a unit.

c   Use the grid on the next page to plot a graph of λ^2 (on the vertical axis) against *M* (on the horizontal axis).

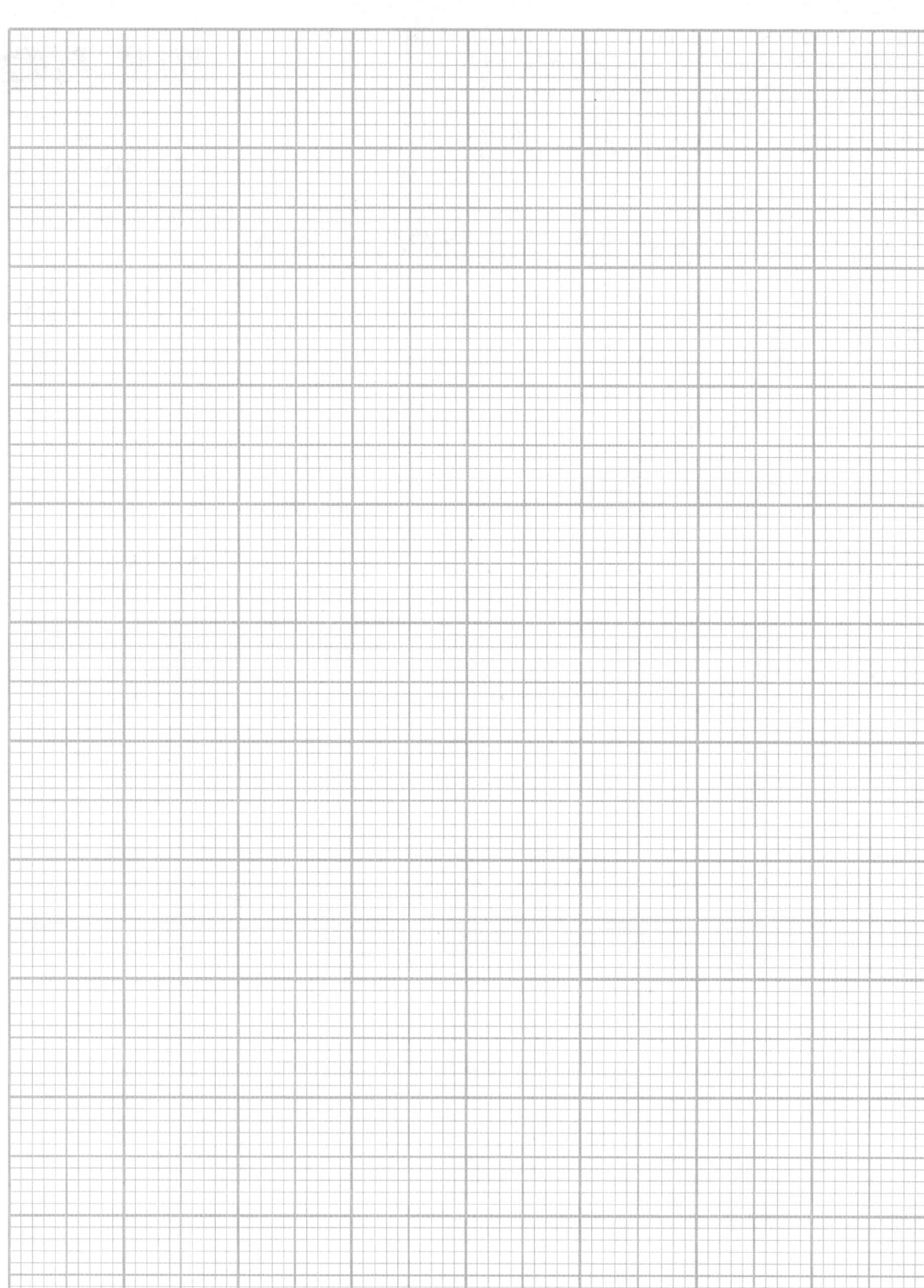

d Draw a straight line of best fit through the points.

e Determine the gradient and intercept of the line.

TIP

Refer to the Skills chapter for advice on choice of scales.

Gradient = ... Intercept = ...

f The equation in part **a** can be applied to change the frequency of the note produced by a guitar string. Which quantity in the equation does the player change to give a different note, and how do they change it?

....

....

g Theory predicts that M and λ^2 are related by:

$$\lambda^2 = \frac{Mg}{\mu f^2}$$

where g is equal to 9.81 m s^{-2}

f is the frequency of the a.c. power supply (written on the power supply unit)

μ is the mass per unit length of the wire

Use your value of the graph gradient to calculate the value of μ, including its unit.

TIP

Check that the values you use have compatible units.

μ = ...

Practical investigation 8.2: Inverse-square law for waves from a point source

Waves from a point source of light spread their energy in all directions so the energy arriving per unit area gets less as the distance from the source increases. Light energy arriving per unit area is called illuminance, and has the unit lux.

In this experiment the illuminance is found using a light-dependent resistor (LDR) and the data is used to test the theoretical relationship between light energy and distance from the point source.

YOU WILL NEED

Equipment:
• small filament lamp mounted inside a black paper tube • extra small filament lamp • power supply • two connecting leads • LDR mounted on the end of a half-metre rule • ohmmeter • digital callipers.

Safety considerations

• Make sure you have read the Safety advice at the beginning of this book and listen to any advice from your teacher before carrying out this investigation.

• The filament lamps have glass domes and should be handled with care. If they break they could cause cuts.

Method

1 Assemble the two parts of the apparatus as shown in Figure 8.3.

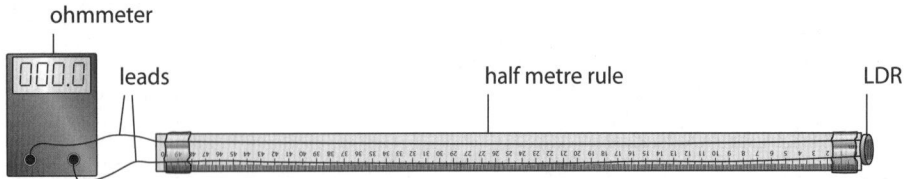

Figure 8.3: Two diagrams showing lamp in tube and wire on metre rule attached to ohmmeter.

---

2 Push the half-metre rule into the paper tube until the LDR touches the glass of the lamp.

Take the reading A on the scale of the rule at the end of the tube, as shown in Figure 8.4.

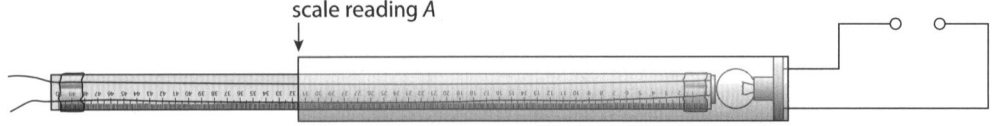

Figure 8.4: Wire on metre rule inside tube with lamp.

Record the value of A in the Results section.

3 Pull the LDR approximately 5 cm away from the lamp and switch on the power supply.

Record the new scale reading B and the ohmmeter reading R in Table 8.2 in the Results section.

4 Pull the LDR further away from the lamp in steps, recording the values of B and R at each step until you have six sets of values in Table 8.2.

5 There is a zero error E due to the distance between the lamp filament and the LDR sensing surface when the LDR touches the glass of the lamp, as shown in Figure 8.5.

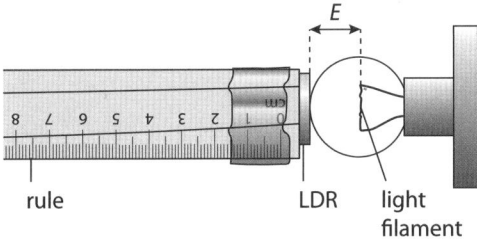

rule   LDR   light
                               filament

Figure 8.5: Close-up of Figure 8.4 at the lamp.

6 Measure the extra lamp and the LDR to find an estimate for E.

Record your value of E in the Results section.

> **TIP**
>
> Keep the ohmmeter range set at 20 kΩ throughout the experiment.

Results

Scale reading A = ... cm

Table 8.2: Results table.

Estimated value for E = ... cm

Analysis, conclusion and evaluation

a Calculate the values of x using $x = A - B + E$ and add them to Table 8.2.

b Calculate the values of $\dfrac{1}{x^2}$ first in cm^{-2} then in m^{-2} and add them to Table 8.2.

c The relationship between illuminance and the LDR resistance is given in the manufacturer's data sheet as the log–log graph shown in Figure 8.6.

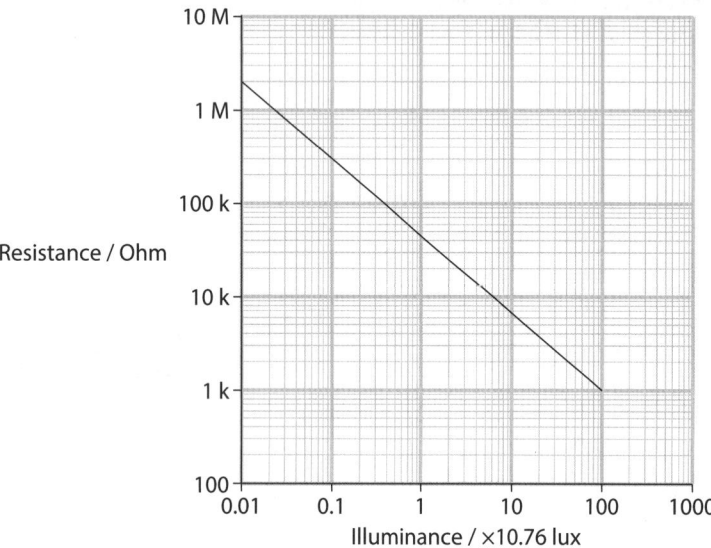

Figure 8.6: The relationship between illuminance and the LDR resistance.

The graph has the equation:

$$L = 10.76 \times \left(\frac{42}{R}\right)^{1.3}$$

where L is in lux and R is in kΩ.

Use the equation to calculate the values of L and add them to Table 8.2.

Ensure that every column in Table 8.2 has a heading with the quantity and units.

d Use the graph grid to plot a graph of L (on the vertical axis) against $\dfrac{1}{x^2}$ (on the horizontal axis).

e   Draw a straight line of best fit through the points.

f   Determine the gradient and intercept of the line.

Gradient = ... Intercept = ...

g   Theory predicts that L is proportional to $\dfrac{1}{x^2}$ (an inverse-square relationship).

Explain whether your graph supports this theory.

....

....

Practical investigation 8.3: Refraction of light waves by a lens

Figure 8.7 is a diagram of light passing through a converging lens.

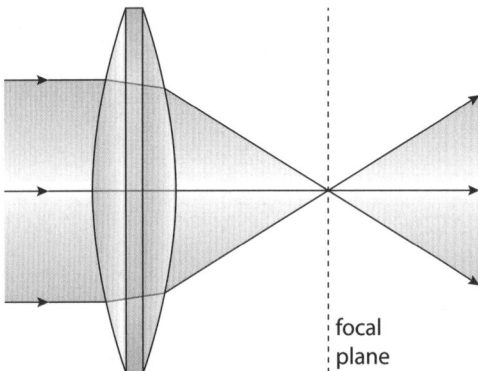

focal plane

Figure 8.7: Light passing through a converging lens.

The speed of light is lower inside the lens material. This causes **refraction** and changes the direction of the waves so that they converge at the focus.

The shape of the lens determines how far the focus is from the lens (the focal length f).

In this investigation you will make measurements which will enable you to calculate the refractive index of the lens material.

YOU WILL NEED

Equipment:
- two converging lenses with different focal lengths • digital callipers
- A4-sized white screen • LED torch • metre rule.

Safety considerations

- Make sure you have read the Safety advice at the beginning of this book and listen to any advice from your teacher before carrying out this investigation.

- There are no special safety issues in this investigation.

Method

1   Select the thinner of the two lenses.

2   Use the digital callipers to measure the diameter D, the thickness T and the edge thickness E as shown in Figure 8.8.

Record the values in Table 8.3 in the Results section.

3   Stand the white screen on the bench with the torch shining on it from a distance of approximately 90 cm. Move the lens along between the torch and the screen until there is a sharp image of the torch LEDs on the screen. Measure the distance u between the torch and the lens and the distance v between the lens and the screen, as shown in Figure 8.9.

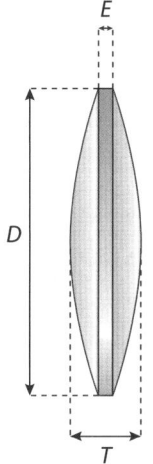

Figure 8.8: Converging lens, edge-on.

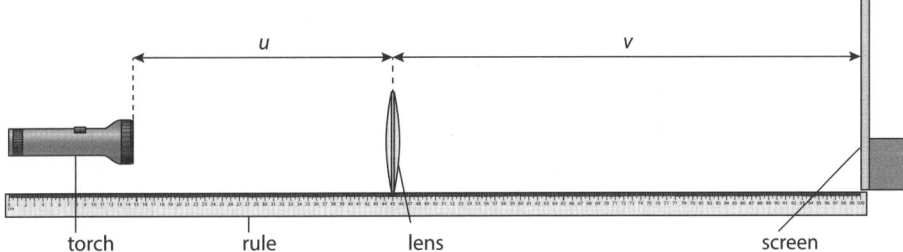

Figure 8.9: Torch, lens, screen and metre rule.

Record the values of u and v in Table 8.4 in the Results section.

4   Repeat steps **2** and **3** using the other lens.

TIP

E may vary around the lens so take the average of several readings. These can be written down in the space next to Table 8.3.

Results

	D / mm	T / mm	E / mm	C / mm	R / mm
Thinner lens					
Thicker lens					

Table 8.3: Results table.

	u / cm	v / cm	f / cm	R / cm	η
Thinner lens					
Thicker lens					

Table 8.4: Results table.

TIP
Note that the values of R are copied to Table 8.4 but the unit is changed.

Analysis, conclusion and evaluation

a   For each row in Table 8.3:

   i   Calculate C using $C = T - E$

   ii   Calculate the radius of curvature R of the lens surfaces using:

   $$R = \frac{D^2 + C^2}{4C}$$

b   For each row in Table 8.4:

   i   Calculate the focal length f of the lens using:

   $$f = \frac{uv}{u + v}$$

   ii   Calculate the refractive index η of the lens material using:

   $$\eta = \frac{R}{2f} + 1$$

c Calculate the percentage difference between the two values of η.

TIP

See treatment of uncertainties in the Skills chapter.

Percentage difference = ...%

d Estimate the percentage uncertainty in your values of u.

Percentage uncertainty = ...%

e Explain whether your answers in parts **c** and **d** support a suggestion that the refractive index η is the same for the two lenses.

....

....

TIP

Compare the difference between your η values with the experimental uncertainty.

f Name the measurement that you found the most difficult and describe the difficulty.

....

....

g Suggest a change in the experimental procedure that could help with the difficulty in part **f**.

....

....

Chapter 9

Planning and data analysis

CHAPTER OUTLINE

This chapter relates to Chapter P2: Planning, analysis and evaluation, in the coursebook.

In this chapter you will complete exercises and investigations on:

- 9.1 Planning data analysis

- 9.2 Treatment of uncertainties

- 9.3 Planning investigation into how the acceleration of a vehicle rolling down an inclined plane varies with the angle of the plane

- 9.4 The acceleration of a vehicle rolling down an inclined plane

- 9.5 Planning investigation into how the current in an LDR varies with the distance from a light source

- 9.6 The resistance of an LDR

- 9.7 Planning investigation into how the electromotive force (e.m.f.) of a photovoltaic cell varies with the thickness of an absorber.

Practical investigation 9.1: Planning data analysis

One of the skills in planning an investigation involves identifying the **independent variable**, the **dependent variable** and the variables to be **controlled**. Another skill is explaining how the data collected will be analysed.

In this exercise you will identify the variables and explain how the data collected will be analysed.

Part 1: Projectile motion

A ball is released from rest so that it travels a distance L along a slope. When it reaches the bottom of the slope it leaves the table horizontally and lands at P, a horizontal distance R from the end of the table, as shown in Figure 9.1.

KEY WORDS

independent variable: the variable in an experiment with a value that is altered by the experimenter

dependent variable: the variable in an experiment with a value that changes as the independent value is altered by the experimenter

control variable: a quantity that has to be kept constant otherwise the relationship between the other variables is not tested fairly

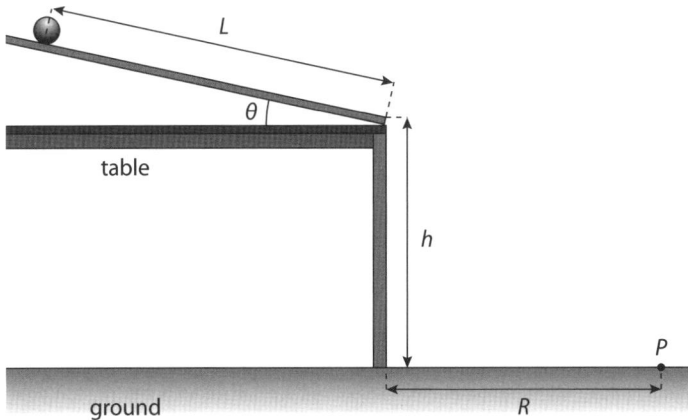

Figure 9.1: Ball on inclined plane on table top.

It is suggested that the relationship between R and L is:

$R = \sqrt{4hL\sin\theta}$

where h and θ are constants.

Variables

List the dependent variable, the independent variable and the variables that should be controlled. The variables to be controlled are quantities that must be kept constant.

*   Dependent variable: ....

*   Independent variable: ....

*   Variables to be controlled: ....

    ....

Data analysis

Describe how you can analyse the data to show the relationship between R and L. Your account should include plotting a graph and using either the gradient or y-intercept of the graph to determine a value for h.

*   Graph to plot: ....

*   Equation of straight-line graph: ....

*   Gradient = ...

* $h = $ ....

Part 2: Current in a filament lamp

The current in a filament lamp is measured for different potential differences across it.

It is suggested that the relationship between I and V is:

$I = pV^q$

where p and q are constants.

Variables

List the dependent variable and the independent variable.

*   Dependent variable: ....

*   Independent variable: ....

Data analysis

Describe how you can analyse the data to show the relationship between I and V. Your account should include plotting a graph and using either the gradient or the y-intercept of the graph to determine values for p and q.

*   Graph to plot: ....

*   Equation of straight-line graph: ....

*   Gradient = ... y-intercept = ...

* p = ... q = ...

Practical investigation 9.2: Treatment of uncertainties

In this exercise you will be given data to enable you to calculate density and determine the **absolute uncertainty** in your answer.

KEY EQUATIONS

absolute uncertainty = gradient of line of best fit − gradient of worst acceptable line

absolute uncertainty = y-intercept of line of best fit − y-intercept of worst acceptable line

Note:

$$\text{density} = \frac{\text{mass}}{\text{volume}}$$

Part 1: Density of a liquid

The mass of a beaker is measured using a top-pan balance measuring to the nearest gram.

A volume of cooking oil is measured using a measuring cylinder.

Results

Mass of empty beaker = (324 ± 1) g

Volume of cooking oil added to beaker = (300 ± 5) ml

Mass of beaker and cooking oil = (603 ± 1) g

Data analysis

a   Calculate the mass of cooking oil. Include the absolute uncertainty.

Mass = ... ± .... g

b   Calculate the maximum and minimum values of the mass of the cooking oil using the results provided.

Maximum mass = ... g   Minimum mass = ... g

c   Determine the absolute uncertainty in the mass of cooking oil using your answer to part b.

Absolute uncertainty in the mass = ± .... g

d   Calculate the density of cooking oil.

Density = ... g cm^{-3}

e   Calculate the percentage uncertainty in the mass of the cooking oil.

Percentage uncertainty = ...%

f   Calculate the percentage uncertainty in the volume of the cooking oil.

Percentage uncertainty = ...%

g   Calculate the percentage uncertainty in the density of the cooking oil.

Percentage uncertainty = ...%

h   Determine the absolute uncertainty in the density of the cooking oil.

Absolute uncertainty = ...

i   Using your answers to part **b** determine the maximum density of the cooking oil.

Maximum density = ... g cm^{-3}

j   Using your answers to part **b** determine the minimum density of the cooking oil.

Minimum density = ... g cm^{-3}

k   Determine the absolute uncertainty in the density of the cooking oil using your answers to parts **i** and **j**.

Absolute uncertainty = ...

Part 2: Density of a metal sphere

The diameter of a metal sphere is measured several times in different directions using Vernier callipers. The mass of the sphere is measured using a top-pan balance measuring to the nearest gram.

Results

Mass of sphere = (19 ± 1) g

Diameter of sphere = 1.59 cm, 1.63 cm, 1.61 cm and 1.62 cm

Data analysis

Note:
volume of a sphere = $\dfrac{4\pi r^3}{3}$

a   Calculate the mean diameter of the sphere. Include the absolute uncertainty.

Diameter = ... ± .... cm

b   Calculate the volume of the sphere.

Volume = ... cm^3

c   Calculate the percentage uncertainty in the volume.

Percentage uncertainty = ...%

d   Calculate the density of the metal.

Density = ... g cm^{-3}

e   Calculate the percentage uncertainty in the density.

Percentage uncertainty = ...%

f   Determine the absolute uncertainty in the density.

Absolute uncertainty = ± .... g cm^{-3}

g   Write down the density of the metal including the absolute uncertainty.

Density = ... ± .... g cm^{-3}

Practical investigation 9.3: Planning
How the acceleration of a vehicle rolling down an inclined plane varies with the angle of the plane

A toy car is placed on an inclined plane as shown in Figure 9.2 and allowed to travel down the slope from rest. You will measure the acceleration of the vehicle.

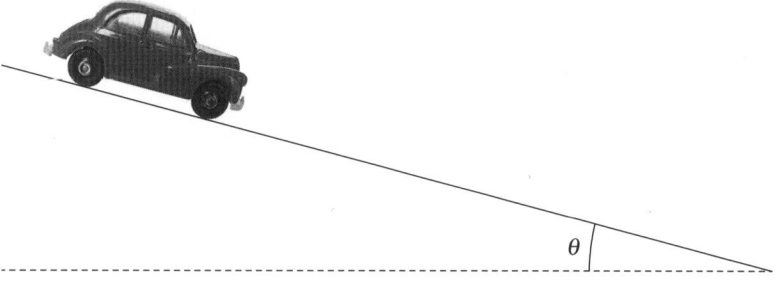

θ

Figure 9.2: Car on slope.

It is suggested that the relationship between the acceleration a and the angle θ of the plane to the horizontal is:

$a = g\sin\theta$

where g is the acceleration due to gravity.

You are going to design a laboratory experiment based on Figure 9.2 to test the relationship between a and θ and explain how a value for g may be determined.

In your account you will:

- write an account of the procedure to be followed
- describe the measurements to be taken
- describe the types of variables involved
- describe how the data is analysed
- give one or two safety precautions that may be taken.

Variables

List the dependent variable, the independent variable and the variables that should be controlled. The variables to be controlled are quantities that must be kept the same.

- Dependent variable: ....
- Independent variable: ....
- Variables to be controlled: ....

....

YOU WILL NEED

List the equipment you will need and draw a labelled diagram of how you will set up the apparatus to obtain the necessary measurements.

- ....
- ....
- ....
- ....
- ....
- ....
- ....
- ....
- ....
- ....

Safety considerations

- Make sure you have read the Safety advice at the beginning of this book and listen to any advice from your teacher before carrying out this investigation.

- ....
- ....

Method

Describe how you will carry out the experiment.

....

....

....

....

....

....

....

....

....

Results

Draw a table of results which can be used to record and process the data from this experiment. You do not have to fill in the table. Remember to include the correct units in the column headings.

Analysis, conclusion and evaluation

a Describe how you can analyse the data to show the relationship between a and θ. Your account should include plotting a graph and using either the gradient or y-intercept of the graph.

....

....

....

....

b Using your knowledge, derive the expression $a = g \sin \theta$.

Practical investigation 9.4: The acceleration of a vehicle rolling down an inclined plane

A trolley (or toy car) is placed on an inclined plane and allowed to travel down the slope from rest. You will measure the time taken for the trolley to travel a constant distance.

YOU WILL NEED

Equipment:
• trolley or toy car • stopwatch • inclined plane • variable support for inclined plane • metre rule • protractor • cardboard box • light gate(s) to assist in the determination of a.

Safety considerations

• Make sure you have read the Safety advice at the beginning of this book and listen to any advice from your teacher before carrying out this investigation.

• Place a cardboard box at the lower end of the inclined plane so that the trolley or toy car does not travel along the floor.

Method

1 Start by measuring the length L of the inclined plane and the distance D that the trolley is going to travel each time as shown in Figure 9.3. Record your readings in the Results section.

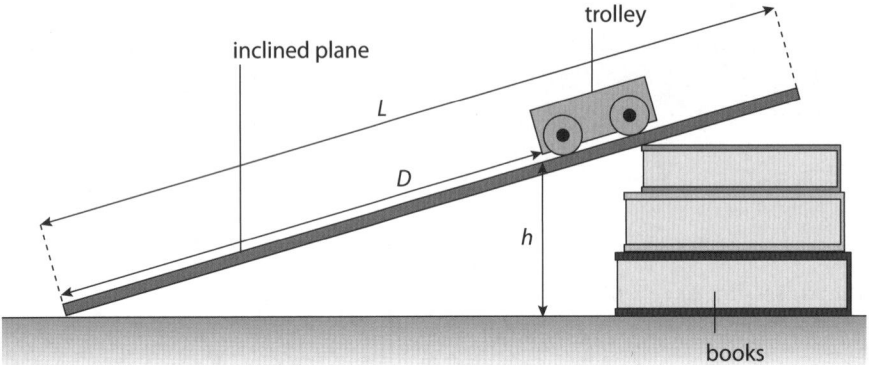

Figure 9.3: Trolley on inclined plane (supported by books).

You will need to keep the distance D constant throughout this experiment, so make a mark on the inclined plane.

2 Measure the height h of the inclined plane at the front of the trolley as shown in Figure 9.3.

3 Place the trolley a distance D from the end of the inclined plane. Release the trolley from rest and measure the time t for the trolley to travel distance D. Repeat the measurement of t for the same height h of the inclined plane. Record your measurements in Table 9.1 in the Results section.

4 Repeat the measurement of the time t using different heights of the inclined plane. Record all your measurements in Table 9.1.

Results

$L =$ .... cm $\qquad D =$ .... cm

h / cm	t / s			
	1st value	2nd value	average	
			±	±
			±	±
			±	±
			±	±
			±	±
			±	±

Table 9.1: Results table.

Analysis, conclusion and evaluation

a Calculate the average value of your two readings for t in Table 9.1. Calculate the uncertainty in each average value of t and add this after $\pm$ in the value for average in the table.

TIP

The uncertainty in t is half the difference between your two readings.

b Calculate the value of $\dfrac{1}{t^2}$ / s^{-2} for all your readings. Use your uncertainty in the average value of t to calculate the absolute uncertainty in $\dfrac{1}{t^2}$. This is done by remembering that the percentage error in $\dfrac{1}{t^2}$ is 2 × the percentage uncertainty in t. Alternatively, you can use your largest and smallest values of $\dfrac{1}{t^2}$ to estimate the uncertainty. Record the absolute uncertainty in each value of $\dfrac{1}{t^2}$ after the $\pm$ symbol. You only need to give a value of this uncertainty to one significant figure.

c Draw a graph on the next page of $\dfrac{1}{t^2}$ / s^{-2} on the y-axis plotted against h on the x-axis.

It is suggested that the relationship between t and L is:

d $\dfrac{2D}{t^2} = \dfrac{gh}{L}$

where g is the acceleration of free fall.
By rearranging the relationship, make $\dfrac{1}{t^2}$ the subject of the formula.

$\dfrac{1}{t^2}$ = ...

e Using your equation from **d**, determine the gradient of your graph of $\dfrac{1}{t^2}$ against h, in terms of g, D, L and the other constants.

Gradient = ...

f Use the uncertainty in your values of $\dfrac{1}{t^2}$ to draw error bars on your graph.

On your graph, draw a straight line of best fit and a worst acceptable straight line.

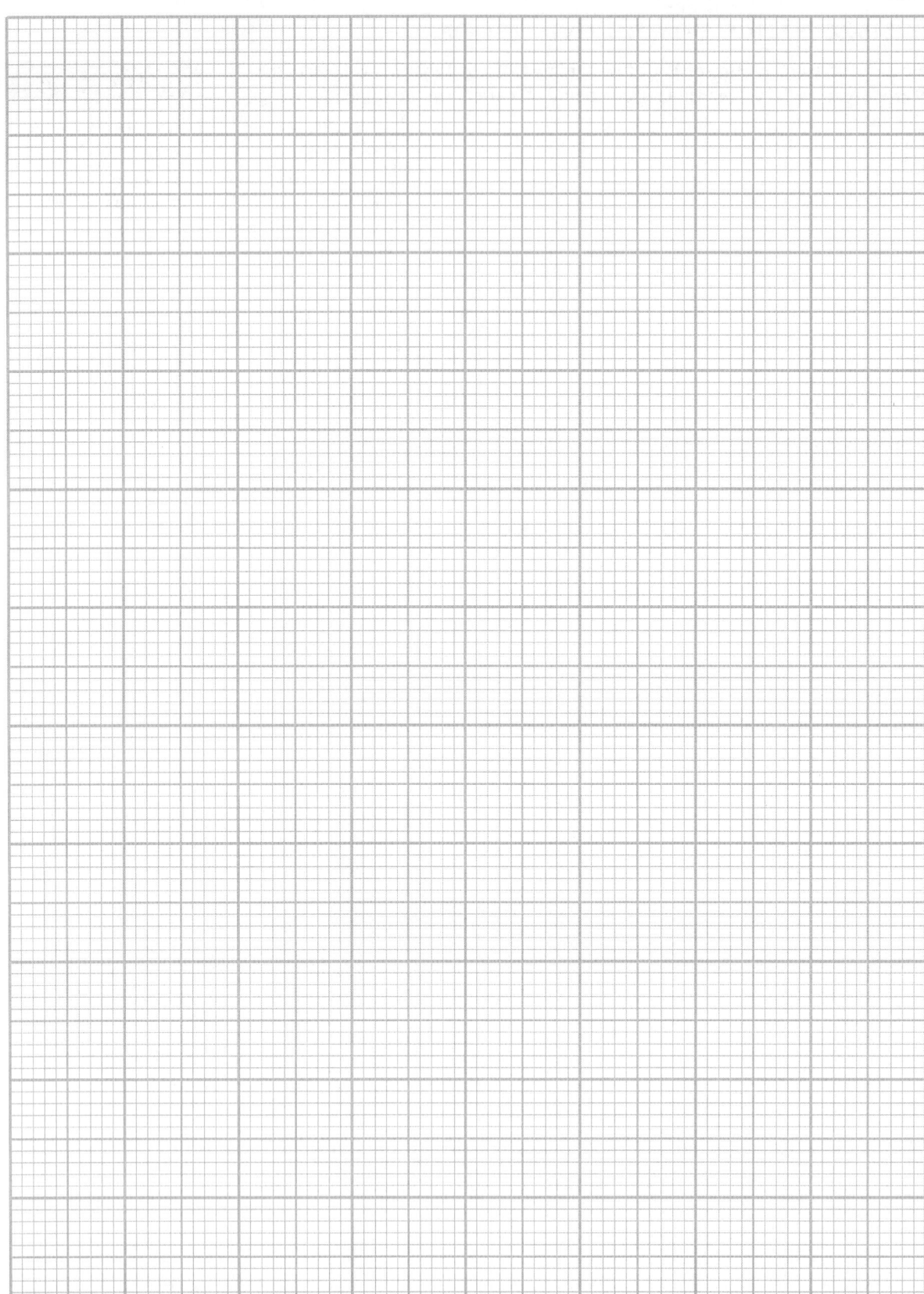

g   Determine the gradient of the line of best fit and the gradient of the line of worst fit. You do not need to give units. Use your value for the uncertainty in the worst-fit line to estimate the uncertainty in your value of the gradient.

Gradient of best-fit line = ...   Gradient of worst-fit line = ...

Uncertainty in gradient = ...

h   Using your value of the gradient and your values of D and L, determine a value for g. Include an appropriate unit for g.

g = ...

i   Determine the percentage uncertainty for g.

Percentage uncertainty = ...%

j   Explain why the graph may have a y-intercept.

....

....

TIP

Think about the forces acting on the trolley.

k   Explain what could be done to reduce the y-intercept.

....

....

Practical investigation 9.5: Planning
How the current in an LDR varies with the distance from a light source

The resistance of a light-dependent resistor (LDR) depends on the intensity of the light falling on the surface of the LDR. One way to investigate the resistance of an LDR is to vary the distance a light source is placed from an LDR and measure the current in the LDR.

Theory suggests that the relationship between the current in an LDR and the distance D at which a light source is placed is:

$I = pD^q$

where p and q are constants.

You are going to design a laboratory experiment to test the relationship between I and D. In your account you will:

- write an account of the procedure to be followed
- describe the measurements to be taken
- describe the types of variables involved
- describe how the data is analysed
- give one or two safety precautions that may be taken.

Variables

List the dependent variable, the independent variable and the variables that should be controlled. The variables to be controlled are quantities that must be kept the same.

- Dependent variable: ....

- Independent variable: ....

- Variables to be controlled: ....

  ....

YOU WILL NEED

List the equipment you will need and draw a labelled diagram of how you will set up the apparatus to obtain the necessary measurements.

- ....
- ....
- ....
- ....
- ....
- ....
- ....
- ....

CONTINUED

Safety considerations

- Make sure you have read the Safety advice at the beginning of this book and listen to any advice from your teacher before carrying out this investigation.

- ....

- ....

Method

Describe how you will carry out the experiment.

....

....

....

....

....

....

....

....

....

....

Results

Draw a table of results which can be used to record and process the data from this experiment. You do not have to fill in the table. Remember to include the correct units in the column headings.

Data analysis

a Describe how you can analyse the data to show the relationship between I and D. Your account should include plotting a graph and using the gradient and intercept of the graph to determine the constants p and q.

....

....

....

....

Practical investigation 9.6:
Data analysis
The resistance of an LDR

The resistance of a light-dependent resistor (LDR) depends on the intensity of the light falling on the surface of the LDR. One way to investigate the resistance of an LDR is to vary the distance a light source is placed from an LDR and measure the resistance of the LDR.

It is suggested that the relationship between the resistance R of an LDR and the distance D at which a light source is placed is:

$R = kD^J$

where J and k are constants.

A similar experiment to the plan you have completed in Practical investigation 9.5 is carried out. The equipment is set up as shown in Figure 9.4. In this experiment, both current and potential difference are measured and recorded for different distances D.

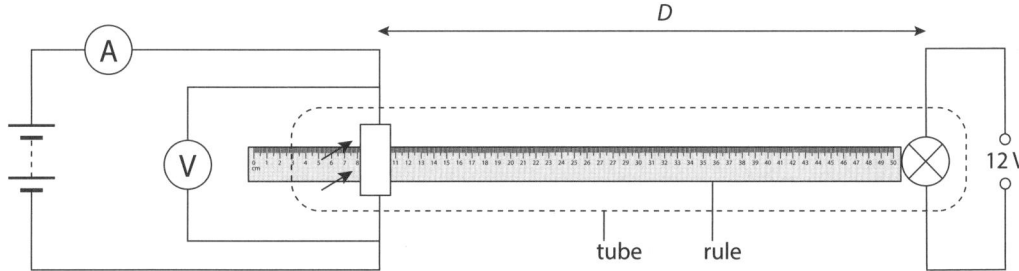

Figure 9.4: Circuit diagram with tube and rule.

Results

The voltmeter reading $V = (5.6 \pm 0.2)$ V

D / cm	I / mA	lg (D / cm)	R / Ω	lg (R / Ω)
22.5	9.6 ± 0.2		±	±
34.5	5.2 ± 0.2		±	±
51.0	3.0 ± 0.2		±	±
64.0	2.2 ± 0.2		±	±
81.5	1.6 ± 0.2		±	±
96.5	1.3 ± 0.2		±	±

Table 9.2: Results table.

Analysis, conclusion and evaluation

a   Calculate the values of R for all the readings. Use the uncertainty in V and the uncertainty in I to calculate the absolute uncertainty in R. Remember that the percentage error in R is the percentage uncertainty in I added to the percentage uncertainty in V. Alternatively you can use your largest and smallest values of I and V to estimate the uncertainty. Remember that:

> **KEY EQUATIONS**
>
> $R_{max} = \dfrac{V_{max}}{I_{min}}$ and $R_{min} = \dfrac{V_{min}}{I_{max}}$

TIP

The number of decimal places in a logarithmic quantity should correspond to the number of significant figures in the raw data.

b Calculate the values of lg (D / cm) and lg (R / Ω) for all the readings. Use your uncertainty in the value of R to calculate the absolute uncertainty in lg (R / Ω). Record the absolute uncertainty in each value of lg (R / Ω) after the ± symbol.

c Draw a graph of lg (R / Ω) on the y-axis plotted against lg (D / cm) on the x-axis using the grid on the next page.

d It is suggested that the relationship between R and D is:

$R = kD^J$

where J and k are constants.

By rearranging the relationship, make lg R the subject of the formula.

lg R = ...

e Using your equation from part **d**, determine the gradient and y-intercept of your graph of lg (R / Ω) against lg (D / cm), in terms of J and k.

Gradient = ... y-intercept = ...

f Use the uncertainty in your values of lg R to draw error bars on your graph. On your graph, draw a straight line of best fit and a worst acceptable straight line.

g Determine the gradient of the line of best fit and the gradient of the line of worst fit. You do not need to give units. Use your value for the uncertainty in the worst-fit line to estimate the uncertainty in your value of the gradient.

Gradient of best-fit line = ... Gradient of worst-fit line = ...

Uncertainty in gradient = ...

h Determine the y-intercept of the line of best fit and the y-intercept of the line of worst fit. You do not need to give units. Use your value for the uncertainty in the worst-fit line to estimate the uncertainty in your value of the y-intercept.

y-intercept of best-fit line = ... y-intercept of worst-fit line = ...
Uncertainty in y-intercept = ...

i   Using your values of the gradient and y-intercept, determine values for J and k. Do *not* include units.

$J = $ .... $k = $ ....

j   Determine the absolute uncertainties in J and k.

Absolute uncertainty in $J = $ ....   Absolute uncertainty in $k = $ ....

k   Explain the limitations in the measurements of D and R.

....

....

> **TIP**
>
> Think about how D was measured.

l   Explain how these measurements could be improved.

....

....

m   Explain the likely effect on the graph and the values of J and k of the limitations and improvements in your answers to parts **k** and **l**.

....

....

> **TIP**
>
> What will be the impact on both the gradient and the y-intercept?

Practical investigation 9.7: Planning
How the electromotive force (e.m.f.) of a photovoltaic cell varies with the thickness of an absorber

A photovoltaic cell generates an electromotive force (e.m.f.) when light falls on the cell. When a plastic slide is placed between the lamp and the cell, the e.m.f. is reduced.

A student is investigating how the number n of plastic slides affects the e.m.f. E.

A student suggests that:

$E = E_0 e^{-nkt}$

where E_0 is the e.m.f. without plastic inserted, t is the thickness of one slide and k is a constant.

You are going to design a laboratory experiment to test the relationship between E and t. In your account you will:

- write an account of the procedure to be followed
- describe the measurements to be taken
- describe the types of variables involved
- describe how the data is analysed to obtain values for k and E_0
- give one or two safety precautions that may be taken.

Variables

List the dependent variable, the independent variable and the variables that should be controlled. The variables to be controlled are quantities that must be kept the same.

- Dependent variable: ....

- Independent variable: ....

- Variables to be controlled: ....

  ....

YOU WILL NEED

List the equipment you will need and draw a labelled diagram of how you will set up the apparatus to obtain the necessary measurements.

- ....
- ....
- ....
- ....
- ....
- ....
- ....
- ....
- ....
- ....

CONTINUED

Safety considerations

- Make sure you have read the Safety advice at the beginning of this book and listen to any advice from your teacher before carrying out this investigation.

- ....

- ....

Method

Describe how you will carry out the experiment.

....

....

....

....

....

....

....

....

....

....

Results

Draw a table of results which can be used to record and process the data from this experiment. You do not have to fill in the table. Remember to include the correct units in the column headings.

Analysis, conclusion and evaluation

a   Describe how you can analyse the data to test the relationship between E and t. Your account should include plotting a graph and using the gradient and y-intercept of the graph to determine values for E_0 and k.

   ....

   ....

   ....

   ....

Circular motion and gravitational fields

Practical investigation 10.1: Circular motion

The **centripetal acceleration** of a mass moving at constant speed in a circle depends on the radius of the circle and on the angular speed of the object. In this practical you will investigate this relationship and confirm the theoretical formula for centripetal acceleration.

KEY WORDS

centripetal acceleration: the acceleration of an object towards the centre of its circular motion

YOU WILL NEED

Equipment:
- short length of tube • stopwatch • 100 g or 50 g masses or 10 g washers
- string (about 1 m) • rubber bung with hole • metre rule • marker pen.

Safety considerations

- Make sure you have read the Safety advice at the beginning of this book and listen to any advice from your teacher before carrying out this investigation.

- Wear eye protection during the experiment.

- Make sure that there is enough space for you to rotate the bung without any risk to other people or apparatus.

Method

1   Set up the apparatus as shown in Figure 10.1.

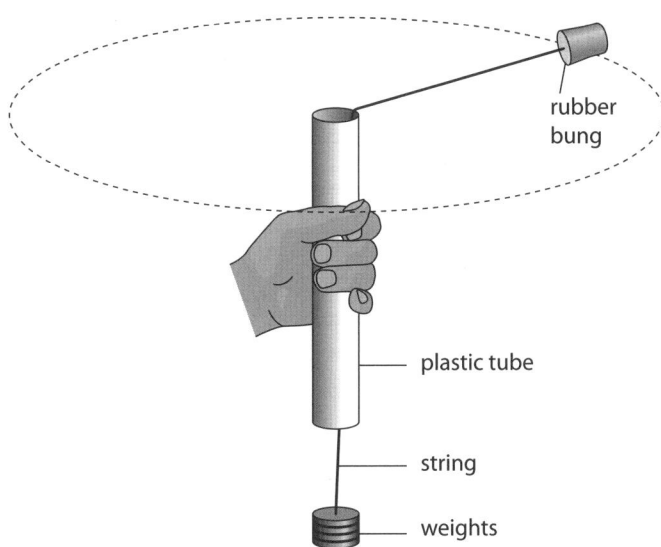

Figure 10.1: Rubber bung, attached by string to weight, through tube in motion.

2   Start by making an initial trial in which you hold the tube and whirl the bung around your head, keeping the radius of the orbit constant. You need to make sure that the string can move freely up and down the tube and that the weights do not rise to touch the bottom of the tube. The weights then provide the resultant force that causes centripetal acceleration acting on the rubber bung to pull it in a circle.

3   You may find it helpful to start with a hanging weight whose mass is about three times the mass of the rubber bung and with a radius of about 70 cm; but you can choose values which enable you to produce a reasonable horizontal circular motion of the bung.

4   You need to keep the radius of the circle constant throughout this experiment, so make a mark on the string as it enters the tube. You can then adjust the frequency of your rotation to make sure that this mark is in the same position each time. You will need to practise keeping the mark in the same position while whirling the bung in a circle.

5   With the help of another student, measure the time for 10 complete oscillations. Repeat and average your results. You may need to make several repeats.

6   Repeat the measurement of the time for 10 oscillations using different weights hanging from the end of the string. You may know that the mass of each weight is, for example, 100 g (0.100 kg). Otherwise, you should measure the mass m of the weights. Record all your readings for the mass m of the weights and the time for 10 oscillations T_{10} in Table 10.1 in the Results section.

7   Measure the radius R of the orbit of the rubber bung. This should be measured from the centre of the rubber bung to the centre of the tube. Record your reading in the space in the Results section.

Results

| m / kg | Time for 10 oscillations T_{10} / s | | | T / s | T^{-2} / s^{-2} |
	1st value	2nd value	average		
				±	
				±	
				±	
				±	
				±	
				±	

Table 10.1: Results table.

R = ... m

Analysis, conclusion and evaluation

a   Calculate the average value of your two readings for T_{10} and the period T of one rotation of the rubber bung for each of your readings and record the values in Table 10.1. Calculate the uncertainty in each value of T and add this after ± in the value for T in the table.

> **TIP**
>
> The uncertainty in T_{10} is half the difference in your two readings.
>
> The uncertainty in T is $\frac{1}{10}$th of the uncertainty in T_{10}.

b   Calculate the value of T^{-2} / s^{-2} for all of your readings. Use your uncertainty in T to calculate the absolute uncertainty in T^{-2}. This is done by remembering that the percentage error in T^{-2} is 2 × the percentage error in T. Alternatively, you can use your largest and smallest values of T^{-2} to estimate the uncertainty. Record the absolute uncertainty in each value of T^{-2} after the ± symbol. You only need to give a value of this uncertainty to one significant figure.

> **TIP**
>
> T^{-2} is the same as $\frac{1}{T^2}$.

c   Draw a graph of T^{-2} / s^{-2} on the y-axis plotted against m / kg on the x-axis using the grid on the next page.

Theory suggests that the **resultant force** that causes centripetal acceleration is given by the equation:

$$F = MR\omega^2 = MR\left(\frac{2\pi}{T}\right)^2$$

where ω is the angular velocity of the rubber bung and M is the mass of the rubber bung.

d Since the force keeping the bung rotating in the circle is the weight mg of the hanging masses:

$$mg = \frac{4\pi^2 M R}{T^2}$$

where g is 9.81 m s^{-2}.

Make T^{-2} the subject of the formula by rearranging this equation.

$T^{-2} = $ ....

e Using your equation from part **d**, determine the gradient of your graph of T^{-2} against m, in terms of g, M, R and the other constants.

Gradient = ...

f On your graph, draw a straight line of best fit. Use the uncertainty in your values of T^{-2} to draw error bars on your graph, then draw a worst acceptable straight line.

g Determine the gradient of the line of best fit and the gradient of the line of worst fit. You do not need to give units. Use your value for the uncertainty in the worst-fit line to estimate the uncertainty in your value of the gradient.

Gradient of best-fit line = ... Gradient of worst-fit line = ...

Uncertainty in gradient = ...

h Using your value of the gradient and your value of R, determine the mass M of the rubber bung and its percentage uncertainty.

Mass of rubber bung = ... kg   Uncertainty = ...%

KEY EQUATIONS

resultant force that causes centripetal acceleration:

= mass × centripetal acceleration;

$$F = Mr\omega^2 = \frac{Mv^2}{r}$$

$$\omega = \frac{2\pi}{T}$$

TIP

Your best-fit line should have roughly equal numbers of your plotted points on either side of the line. Your worst-fit line should, hopefully, go through all the error bars, sometimes above and sometimes below your actual points.

i Although the weight is providing the force that ultimately acts on the rubber bung, what is the name of the force in the string itself?

....

Suggest an effect that friction between the plastic tube and the string has on this experiment.

....

....

....

j Newton's second law was used to obtain the formula used. How does **Newton's third law** apply to this situation?

....

....

....

KEY WORDS
Newton's third law: when two bodies interact, the forces they exert on each other are equal and opposite

Practical investigation 10.2: Planning The conical pendulum

In a conical pendulum, a mass hanging vertically on a string suspended from a fixed point is made to move round in a horizontal circle at a constant angular speed.

A video is taken of a toy aeroplane attached to a string and flying in a horizontal circle. One frame of the video is shown in Figure 10.3, where the toy aeroplane is shown at the extreme edge of the motion. A protractor is superimposed on the video frame.

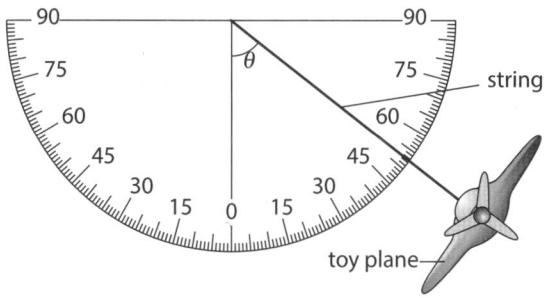

Figure 10.3: Toy plane, protractor, string.

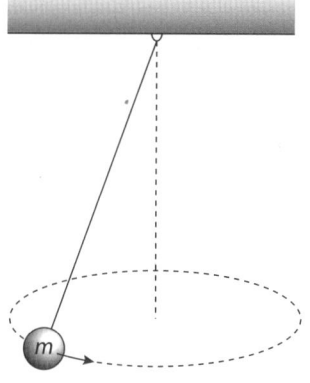

Figure 10.2: Conical pendulum.

The plane can be set to fly at different speeds. This alters the angular velocity ω of the plane and this changes the angle θ. You may be able to see a video of the motion of the toy plane at: Video library for student access or on YouTube (look for 'aeroplane on a string–conical pendulum').

Theory suggests that:

$$\cos\theta = \frac{g}{L\omega^2}$$

where the angle θ is shown in Figure 10.3, ω is the angular velocity of the plane, L is the length of the string and g is the acceleration due to gravity.

You are going to design a laboratory experiment based on Figure 10.3 to test the relationship between θ and ω. In your account you will:

- write an account of the procedure to be followed
- describe the measurements to be taken
- describe the types of variables involved
- describe how the data is analysed
- give one or two safety precautions that may be taken.

Variables

List the dependent variable, the independent variable and the variables that should be controlled. The variables to be controlled are quantities that must be kept the same.

- Dependent variable: ....

- Independent variable: ....

- Variables to be controlled: ....

  ....

YOU WILL NEED

List the equipment you will need and draw a labelled diagram of how you will set up the apparatus in order to obtain the necessary measurements.

- ....
- ....
- ....
- ....
- ....
- ....
- ....
- ....

Safety considerations

- Make sure you have read the Safety advice at the beginning of this book and listen to any advice from your teacher before carrying out this investigation.

- Suggest one safety feature relevant to this experiment.

....

....

Method

Describe how you will carry out the experiment.

....

....

....

....

....

....

....

....

Results

Draw a table of results that can be used to record and process the data from this experiment. You do *not* have to fill in values in the table. Remember to include the correct units in the column headings.

Analysis, conclusion and evaluation

a   Describe how you can analyse the data to show the relationship between θ and ω. Your account should include plotting a graph and using either the gradient or intercept of the graph.

....

....

....

....

....

....

....

....

b   Resolve the tension force T acting on the mass horizontally to prove that

$T = mL\omega^2$. You will need to know that $\sin\theta = \dfrac{r}{L}$. Resolve T vertically to find

another value for T and, by eliminating T, prove that $\cos\theta = \dfrac{g}{L\omega^2}$.

Practical investigation 10.3: Data analysis
Conical pendulum

The investigation described in the introduction to Practical investigation 10.2 is carried out.

The time taken for the toy plane to make 10 complete circuits around the circle is measured, as well as the angle θ, which is the maximum angle shown on the video. The plane is then set to fly faster and the readings are measured again. The readings are shown in Table 10.2. When the measurement of θ is made, the plane may look as though it is at an extreme edge of the motion; however it might not be quite there because the motion is only shown one frame at a time. This means that the angle θ is only measured to the nearest 1°.

Results

θ / °	$\cos \theta$	Time for 10 circuits T_{10} / s	T / s	ω / s^{-1}	$\frac{1}{\omega^2}$ / s^2
10 ± 1	±	14.1			
22 ± 1	±	13.7			
32 ± 1	±	13.1			
42 ± 1	±	12.3			
53 ± 1	±	11.0			
71 ± 1	±	8.1			

Table 10.2: Results table.

Analysis, conclusion and evaluation

a   To verify the relationship $\cos \theta = \dfrac{g}{L\omega^2}$, a graph is plotted of $\cos \theta$ on the y-axis

against $\dfrac{1}{\omega^2}$ on the x-axis. Determine an expression for the gradient in terms

of L and g.

Gradient = ...

b   The period T of a circular motion is related to the angular velocity ω by $T = \dfrac{2\pi}{\omega}$.

Calculate and record values of $\cos \theta$, T, ω and $\dfrac{1}{\omega^2}$ in the table. Include also the absolute uncertainty in $\cos \theta$.

> **TIP**
>
> Find the absolute uncertainty for the first value of $\cos \theta$ by finding the difference between $\cos 11°$ and $\cos 10°$.

> **TIP**
>
> Choose a sensible number of significant figures to match the uncertainty in $\cos \theta$, or three significant figures for T, ω and $\dfrac{1}{\omega^2}$, since T_{10} is only given to three significant figures.

c Plot a graph of $\cos \theta$ on the y-axis against $\dfrac{1}{\omega^2}$ / s^2 on the x-axis. Include error bars for $\cos \theta$.

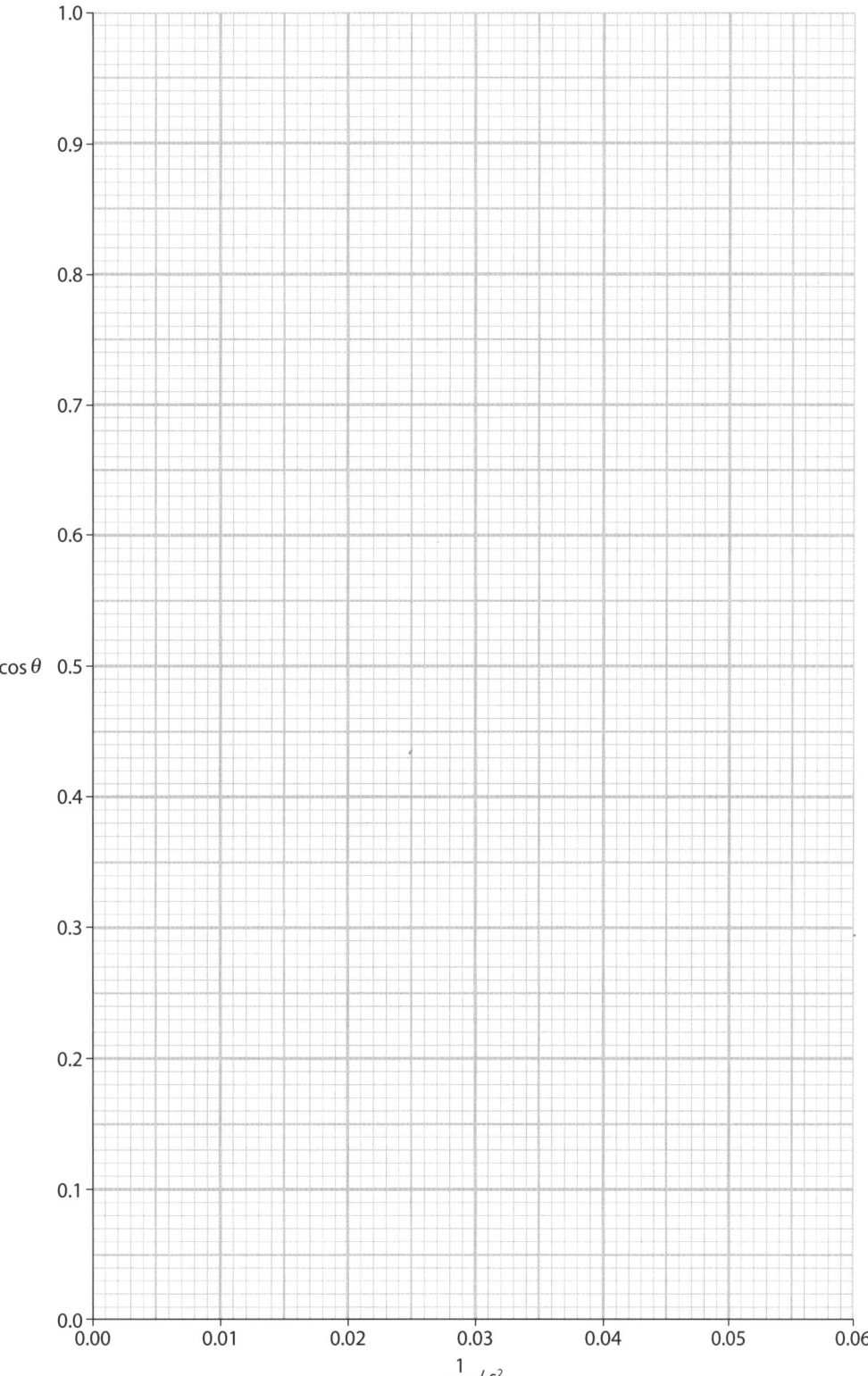

d   Draw the straight line of best fit and a worst acceptable straight line on your graph.

Both lines should be clearly labelled.

e   Determine the gradient of the line of best fit. Include the uncertainty in your answer. Include a unit for your gradient.

Gradient = ...   Uncertainty = ...

f   Using your answers to parts **a** and **e**, determine the value of L. Include the unit of the answer and the percentage uncertainty in L.

Length = ... ± ....%

g   Use your graph to find the value of ω when θ has its smallest possible value and the period of the motion at this value.

....

> **TIP**
>
> Think about the value of $\cos \theta$ when θ is zero.

Practical investigation 10.4:
Data analysis
Planetary motion

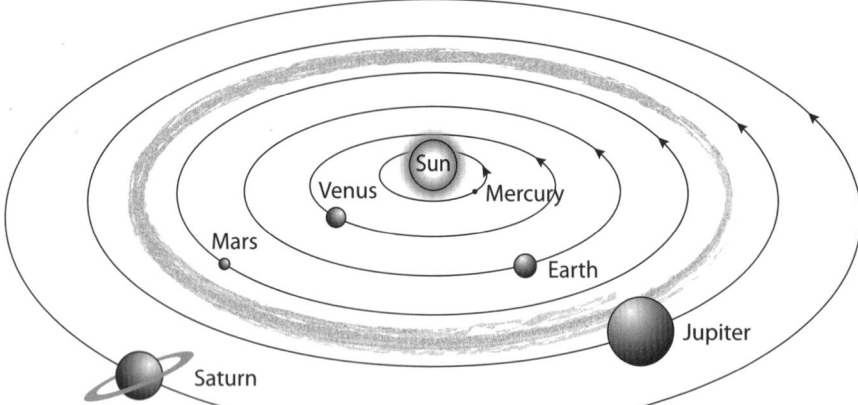

Figure 10.4: Solar system with first six planets.

The period T of the orbit of a planet around the Sun depends upon its distance R from the Sun. In about 1600, Kepler analysed the motion of the planets. You will use some of the data that Kepler used, using modern units, to confirm his relationship between T and R. The uncertainties you will use in the value of R are the differences between two values of R obtained by two different astronomers at about that time. The uncertainties in T are much smaller in comparison and are ignored.

Newton introduced the **law of gravitational attraction** between masses, which explains the relationship between T and R. You will use your graph to determine the mass of the Sun.

The range of distances and times is large and so a logarithmic plot is needed. You will plot best-fit and worst-fit lines using error bars.

> **KEY WORDS**
>
> **Newton's law of gravitation:** any two point masses attract each other with a force that is directly proportional to the product of their mass and inversely proportional to the square of their separation

Results

Values of R and T are given in Table 10.3.

Planet	$R / 10^{10}$ m	$T / 10^6$ s	lg (R / m)	Maximum value of lg (R / m)	lg (T / s)
Mercury	5.8 ± 0.4	7.59	10.763	10.792	6.880
Venus	10.8 ± 0.1	19.4			
Earth	15.0 ± 0.0	31.5			
Mars	22.8 ± 0.1	59.3			
Jupiter	78 ± 3	374			
Saturn	140 ± 40	929			

Table 10.3: Results table.

The logs are all to base 10.

> **TIP**
>
> In the column headed $R / 10^{10}$ m the number 5.8 means that $R = 5.8 \times 10^{10}$ m.

Analysis, conclusion and evaluation

a   Calculate the values of lg (R / m) and the maximum values of lg (R / m) and lg (T / s) for all the planets in the table. One value has been filled in for you. When calculating lg (R / m) to base 10, the number before the decimal point is just the power of 10 that exists in the value of R. To keep the same precision in the data, which is three significant figures, you must keep three decimal places in the logarithm.

b   Plot a graph of lg (T / s) on the y-axis against lg (R / m) on the x-axis. Use your maximum value of lg (R / m) to plot an error bar for all of the planets except Earth.

> **TIP**
>
> Give logs to three decimal places.

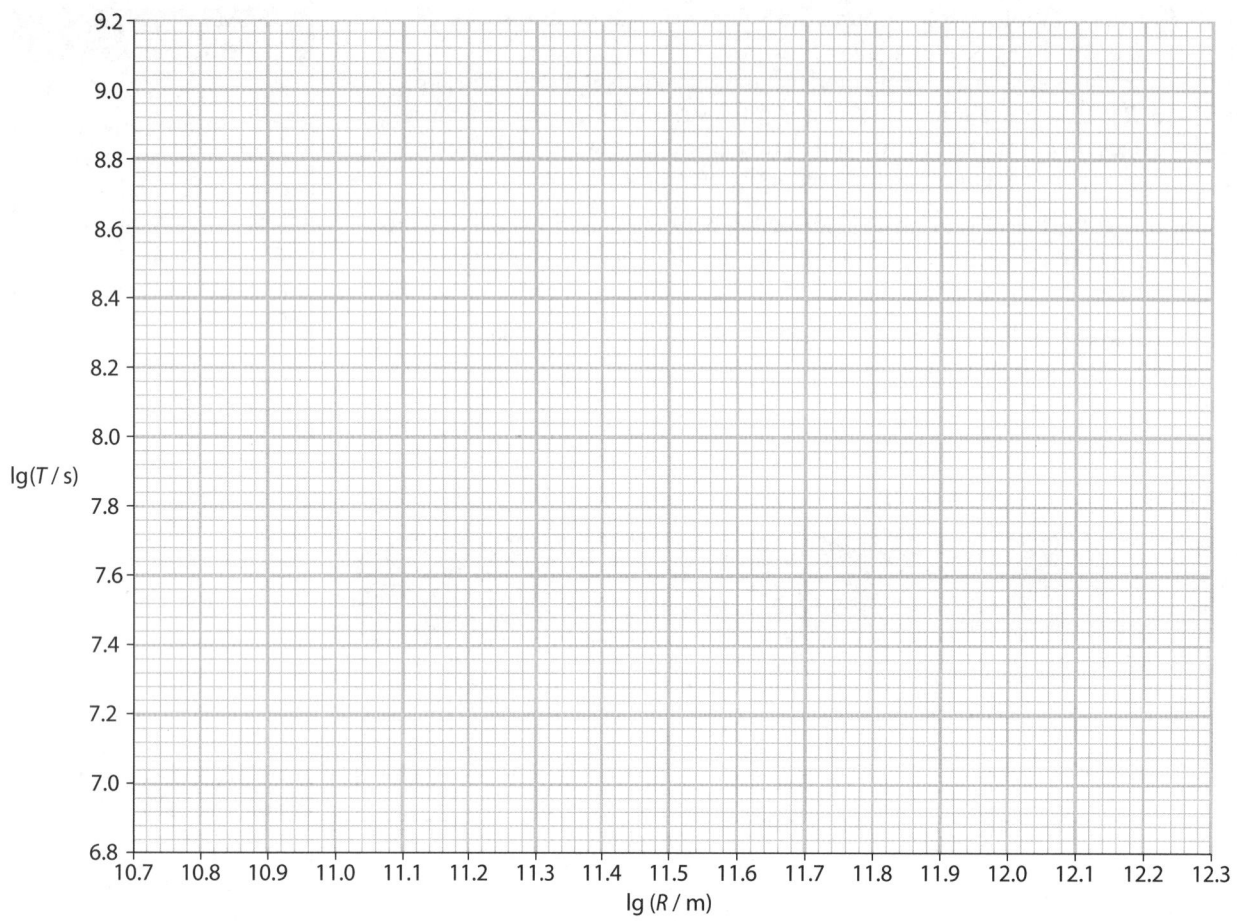

c Draw the straight line of best fit and a worst acceptable straight line on your graph.

d Determine the gradients of the line of best fit and the line of worst fit.

> **TIP**
>
> On your graph show how you obtained the gradient.

Gradient of line of best fit = ... Gradient of line of worst fit = ...

e Calculate the uncertainty in your value for the gradient.

Uncertainty in gradient = ...

f It is suggested that T is related to R by the formula $T = kR^n$, where k and n are constants.

Complete the formula for lg T in terms of R, k and n.

> **TIP**
>
> Take logs to base 10 on both sides of the equation.

lg T = ...

g   Use the gradient of your graph to find a value for n and its uncertainty.

$n = \dots\dots\dots\dots \pm \dots\dots\dots\dots$

h   Using your value for n and the point for Earth on your graph, calculate a value for k. There is no need to calculate an uncertainty.

$k = \dots\dots\dots\dots$

i   Newton's law of gravitation and the formula for centripetal acceleration can be used to show that:

$$T^2 = \frac{4\pi^2 R^3}{GM_s}$$

where G is the universal constant of gravitation and M_S is the mass of the Sun.

Take logarithms to base 10 of both sides of the equation to complete the expression for $2\lg T$.

$2\lg T = \dots\dots\dots \times \lg R + \dots\dots\dots$

j   Explain whether your value for n in part g and its uncertainty is consistent with your equation in part **i**.

....

....

k   Use your value of k from part **h** to calculate the value of the mass of the Sun M_S given that $G = 6.67 \times 10^{-11}$ N m^2 kg^{-2}.

$M_S = \dots\dots\dots$ kg

Practical investigation 10.5:
Data analysis
Gravitational potential

To travel into space (see Figure 10.5) requires a great deal of energy. On return to Earth, for a large part of its journey a rocket is in free fall, accelerating under the influence of Earth's gravity alone. During this time, automated measurements (telemetry) from the Earth establish not only the distance to the rocket but also its speed.

In this analysis you will use the readings of speed from Apollo 11 as it descended towards Earth to check that the formula for **gravitational potential** near the Earth is correct. The readings are shown in Table 10.4. You will also use the data to calculate the mass of the Earth.

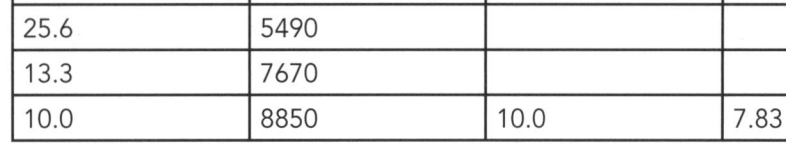

Figure 10.5: Saturn 5 taking off.

R / 10^6 m	v / m s^{-1}	$\dfrac{1}{R}$ / 10^{-8} m^{-1}	v^2 / 10^7 m^2 s^{-2}
242	1520	0.414	0.231
95.2	2720		
54.3	3700		
25.6	5490		
13.3	7670		
10.0	8850	10.0	7.83

Table 10.4: Results table.

> **KEY WORDS**
>
> **gravitational potential:** the work done per unit mass in bringing a mass from infinity to a point

You will need to use this theory:

The gravitational potential ϕ at a distance R from a point mass M is given by the equation

$$\phi = -\frac{GM}{R}$$

where G is the universal constant of gravitation = 6.67×10^{-11} N m^2 kg^{-2}.

The potential energy of a mass m is given by

$$m\phi = -\frac{GMm}{R}$$

The total kinetic and potential energy of the rocket is constant as it falls towards Earth in free fall outside the atmosphere.

This means that at any distance R from the centre of the Earth

$$\frac{1}{2}mv^2 - \frac{GMm}{R} = E$$

where E is a constant, the total energy of the rocket.

> **KEY EQUATION**
>
> **gravitational potential**
>
> $$= \frac{\text{work done}}{\text{mass}}$$
>
> $$\phi = \frac{W}{m}$$

Or more simply

$$v^2 = \frac{2GM}{R} + \frac{2E}{m}$$

You are to test the relationship between v and R and use it to find the mass of the Earth M.

Analysis, conclusion and evaluation

a   If a graph is plotted of v^2 on the y-axis against $\dfrac{1}{R}$ on the x-axis, state the gradient and intercept of the graph, using the symbols G, M, m and E.

Gradient = ...   Intercept = ...

> **TIP**
>
> The powers of 10 in the table can be difficult to understand. The first value of $\dfrac{1}{R}$ is really 0.414×10^{-8}.

b   Values of R and v are given in Table 10.4.

The largest value of R in the table is 241.6×10^6 m. This gives $\dfrac{1}{R} = 0.414 \times 10^{-8}$ m^{-1}. The speed at this distance is 1521 m s^{-1} and this gives $v^2 = 2\,313\,441 = 0.231 \times 10^7$ m^2 s^{-2}.

Check that these values of $\dfrac{1}{R}$ and of v^2 are correct and fill in all of the blank cells in Table 10.4. Make sure that both columns increase down the table.

c Plot a graph of $v^2 / 10^7$ m^2 s^{-2} on the y-axis against $\frac{1}{R}$ / 10^{-8} m^{-1} on the x-axis using the grid provided.

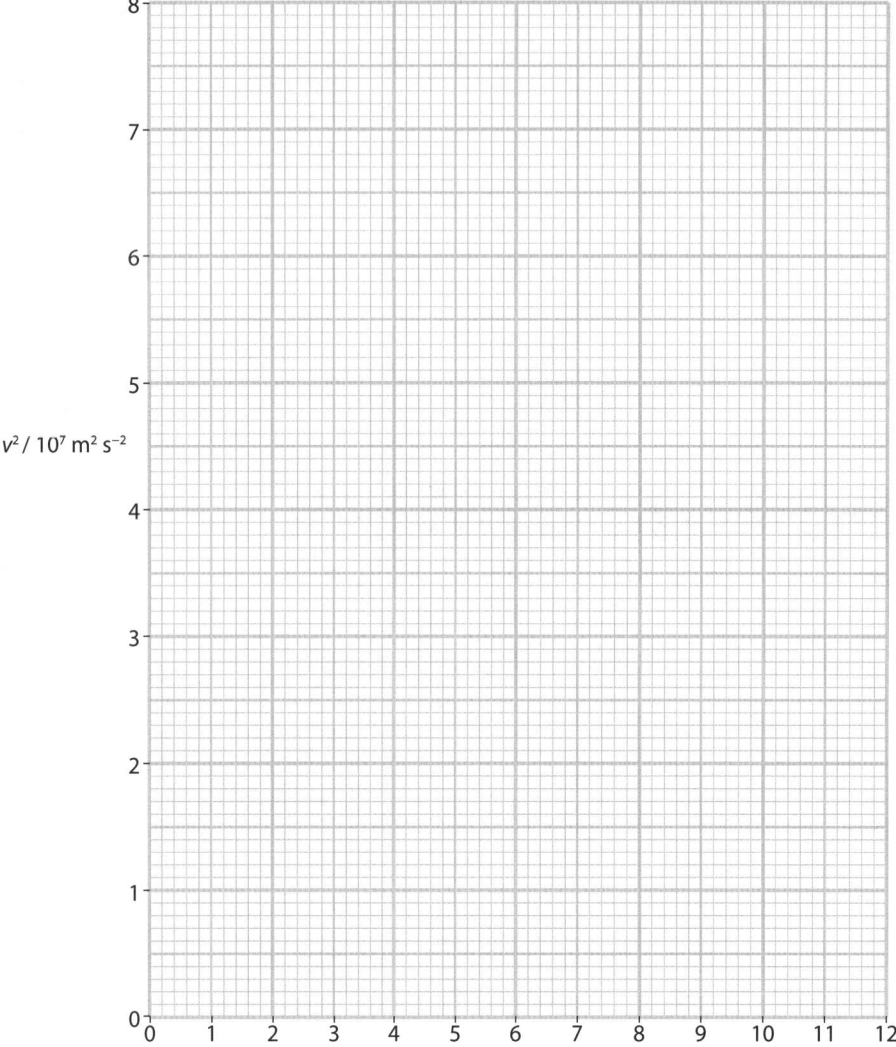

d Determine the gradient of the line of best fit.

Gradient of line of best fit = ...

e Use your value for the gradient in part **d** to determine the mass of the Earth. The value for G is given in the introduction.

Mass of the Earth = ...

> **TIP**
>
> PE is zero at infinity and is negative near the Earth.

f   Determine the intercept on your graph.

Intercept = ...

g   You should have found that the intercept is slightly negative. Using your expression for the intercept in part **f**, suggest why this is possible.

....

....

....

....

Oscillations

Practical investigation 11.1: Oscillation of a metre rule as a pendulum

In this investigation you will investigate the oscillation of a metre rule acting as a pendulum. The period of oscillation depends on the distance between the point of suspension of the rule and its centre of mass.

YOU WILL NEED

Equipment:
• a metre rule with holes drilled at various distances from one end, e.g. 20 cm, 25 cm, 30 cm, 35 cm and 40 cm • stopwatch • a pin held securely in a retort stand with a boss and clamp (possibly with a G-clamp or heavy weight to ensure stability).

Safety considerations

- Ensure that the pin does not enter your eye. In particular, as you bend down to make any adjustments the pin should never become close to your eyes. Safety glasses can be worn or the end of the pin covered with a ball of modelling clay or rubber.

- If the stand is not stable or is liable to topple it can be clamped to the bench or a heavy weight placed on the base.

Method

1   Set up the apparatus as shown in Figure 11.1, with the pin clamped to the stand, using the hole drilled 40 cm from the end of the rule. Make sure that the metre rule swings freely. The rule may hang over the end of the bench, if necessary.

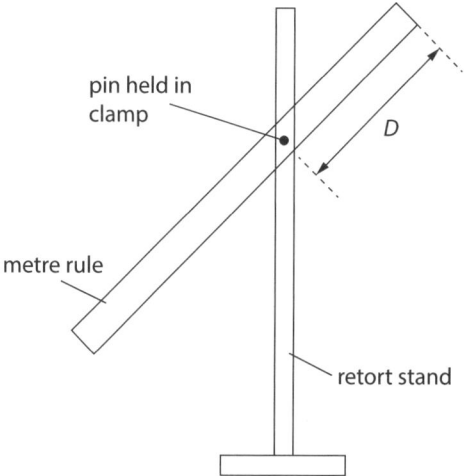

Figure 11.1: Pin through a metre rule (40 cm from one end). Pin attached to a clamp on retort stand.

2   Set the rule into oscillations of small amplitude and take readings to allow you to determine an accurate value of the period T of one oscillation.

You should use a **fiducial mark**. This is a marker to allow you to start and stop your timing at the same point in the oscillation. The best place for the fiducial mark is at the centre of the oscillation, as the pendulum is moving fastest at this point and you should be able to start and stop the stopwatch more accurately. You can place a piece of paper underneath the metre rule as the marker or just use the rod of the retort stand as the marker.

Record your readings in the second column of Table 11.1 in the Results section. Give this column a heading to explain the readings you have taken. Use your readings to calculate the period of oscillation of the metre rule. Measure the distance D from the end of the rule to the pin and include this in the first column of Table 11.1 (the first value $D = 0.400$ m has been done for you).

3   Repeat the procedure, suspending the rule in turn from the other holes in the metre rule. Record your readings in Table 11.1.

TIP

You may choose to record the time for 10 oscillations. To improve accuracy, you can repeat your readings.

KEY WORDS

fiducial mark: a mark or marker used as a point of reference

TIP

One oscillation starts as the metre rule passes the marker going, for example, to the left and finishes the next time it passes the marker going in the same direction.

Results

D / m		period T / s	d / m	T^2d / s^2 m	d^2 / m^2
0.400			0.100		

Table 11.1: Results table.

Analysis, conclusion and evaluation

a   The centre of mass of the metre rule is assumed to be at the 50 cm mark. The distance d between the point of suspension of the rule and its centre of mass is given by

$d = 0.500 - D$

where all values are in metres.

Add values for d in the fourth column in Table 11.1.

b   Theory suggests that T and d are related by the equation

$T^2 = \dfrac{A}{d} + Bd$

where A and B are constants.
Use this equation to show that a graph of T^2d against d^2 is linear.

....

....

....

c   Complete Table 11.1 by calculating values for T^2d and d^2.

d   Draw a graph on the next page of T^2d / s^2 m on the y-axis against d^2 / m^2 on the x-axis. Include the line of best fit.

e   Determine the gradient and the y-intercept of your line of best fit.

Gradient = ... y-intercept = ...

f   Use your answers from part **e** to find the values of A and B. Include the units of your answers.

A = ... unit ....

B = ... unit ....

g   Use the results of your investigation to find a value for T when $d = 45$ cm.

T = ...

Practical investigation 11.2: The period of oscillation of a steel strip

The period of oscillation of a steel strip depends on the mass placed at the end of the strip, the dimensions of the strip and the Young modulus of steel. In this practical you will investigate the relationship between the length of a strip and the period of its oscillation and use the relationship to determine the Young modulus of steel.

You will use a logarithmic graph, plot best-fit and worst-fit lines and determine uncertainties.

YOU WILL NEED

Equipment:
• steel strip • stopwatch • two 50 g masses (or one 100 g mass) • G-clamp and small blocks of wood • 30 cm ruler or metre rule.

Access to:
• micrometer screw gauge or callipers.

Safety considerations

• Make sure you have read the Safety advice at the beginning of this book and listen to any advice from your teacher before carrying out this investigation.

• Take care not to cut yourself on the strip. Ensure that the amplitude of the oscillation is not so large as to allow the masses to break away from the steel strip.

Method

1 Set up the apparatus as shown in Figure 11.2.

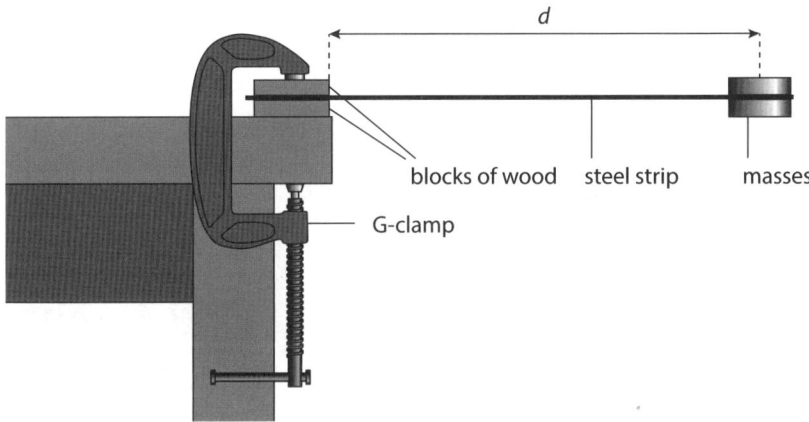

Figure 11.2: Steel strip clamped at one end, mass on the other.

If necessary, fix the masses firmly to the end of the strip and keep them in the same place throughout the investigation. To change the distance d, loosen the G-clamp and move the whole strip, then retighten the G-clamp. Start with d being about half the length of the strip.

2 With the strip held horizontal, measure the distance d from the edge of the wood blocks to the centre of the masses. Record this distance in Table 11.2 in the Results section.

3 Lift the strip upwards a small distance and release it. Record the time T_{10} for 10 complete oscillations of the strip. Repeat this reading and find the average value of T_{10}. Use your value to determine the time T for one oscillation. Record your results in the first row of Table 11.2.

4 Change the value of d and repeat the readings in steps **2** and **3**. Choose five further values of d between 0.120 m and 0.260 m. Record all of your readings in Table 11.2.

5 Measure the width w and the thickness t of the steel strip as accurately as you can. Write these measurements in the Results section.

Results

d / m	T_{10} / s			T / s	lg (d / m)	lg (T / s)
	1st	2nd	average			
				±		±
				±		±
				±		±
				±		±
				±		±
				±		±

Table 11.2: Results table.

Width of strip w = ... Thickness of strip t = ...

Analysis, conclusion and evaluation

a   Using your two values for each value of T_{10}, calculate the uncertainty in each value of T and record your value of this uncertainty where there is a ± sign in Table 11.2.

> **TIP**
>
> The uncertainty is $\frac{1}{2}$ the difference between the largest and smallest values, in this case your two values.

b   Theory suggests that T and d are related by the equation:

$T = kd^n$

where k and n are constants.

Write down an expression for $\lg T$ in terms of k, d and n.

$\lg T$ = ...

c   A graph can be drawn of $\lg (T / s)$ on the y-axis against $\lg (d / m)$ on the x-axis. Use your expression from part **b** to determine the gradient and the y-intercept of this graph in terms of k and n.

Gradient = ... Intercept = ...

d Add your values for lg (d / m) and lg (T / s) to Table 11.2. Include the absolute uncertainties for lg (T / s) by working out the maximum value of lg (T / s) and subtracting from the value of lg (T / s) in the table.

TIP

Include error bars for lg (T / s).
You may assume that the uncertainty in d is negligible.

e Plot a graph of lg (T / s) on the y-axis against lg (d / m) on the x-axis.

f Draw a straight line of best fit and a worst acceptable straight line on your graph.

g Determine the gradient of the line of best fit and the gradient of the line of worst fit. You do not need to give units.

> **TIP**
>
> The uncertainty in the gradient is the difference between the two gradients.

Gradient of best-fit line = ...

Gradient of worst-fit line = ...

h Determine the value of n and its uncertainty shown by your results.

n = ...   Uncertainty in n = ...

i Use a point on your graph and your value of n in part **h** to find the value of $\lg k$ and k in the equation $T = kd^n$. There is no need to give the unit.

Point on graph: $\lg (T / \text{s})$ = ... $\lg (d / \text{m})$ = ...

$\lg k$ = ... k = ...

j Theory suggests that

$$k = \sqrt{\frac{16\pi^2 M}{Ewt^3}}$$

where M is the mass added to the end of the strip, E is the Young modulus, w is the width of the strip and t is its thickness.

Determine a value for E using your value of k, the value of M as 0.100 kg and your values for w and t. If all your values are in SI units, such as m and s, then the unit of E is N m^{-2}.

E = ... N m^{-2}

Practical investigation 11.3: Planning Damped oscillations

This is a planning exercise that uses simple apparatus to check a suggested formula for the exponential decay of an oscillator.

Apparatus may be available for you to perform the experiment after you have written out your plan. If you do use apparatus or take readings, you must obtain the approval of your teacher in advance. In any case, write out your plan before taking any readings.

When a mass swings from side to side, the amplitude of the oscillation slowly decreases. The amplitude is the largest distance of the mass from the equilibrium position during each swing, as shown in Figure 11.3.

KEY WORDS

damped oscillation: an oscillation in which frictional forces cause the energy of the system to be transferred to the surroundings

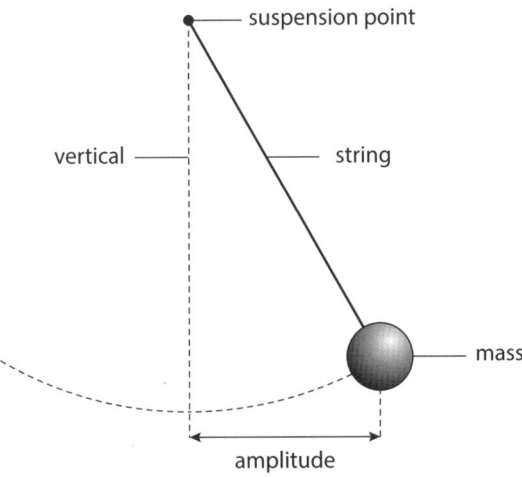

Figure 11.3: Pendulum.

It is suggested that the amplitude A decreases exponentially with the number of swings n according to the equation

$A = A_0 e^{-\lambda n}$

where A_0 is the initial starting amplitude and λ is a constant.

It is further suggested that the value of λ increases if there is more air resistance.

You are going to:

- design a laboratory experiment to test the relationship between A and n

- explain how your results can be used to determine a value for λ

- suggest how the experiment can be used to show that λ increases if air resistance increases.

TIP

Is it easier to measure the amplitude after a certain number of swings or the number of swings after a certain amplitude? Think also about the graph to plot.

Variables

List the dependent variable, the independent variable and the variables that should be controlled in the first part of the experiment to show the relationship between the amplitude *A* and the number of swings *n*. The variables to be controlled are the quantities that must be kept the same.

*   Dependent variable: ....

*   Independent variable: ....

*   Variables to be controlled: ....

    ....

YOU WILL NEED

It is not easy to judge when a mass has reached the furthest distance from the centre. Think about how you will measure the dependent and independent variables. What apparatus will you need? How will it be arranged? How will you know that it is in the correct position to measure the amplitude? The choice of apparatus is yours, so think carefully.

List the equipment required here.

*   ....

*   ....

*   ....

*   ....

*   ....

*   ....

Draw a diagram of how to set up the apparatus.

Safety considerations

State **one** precaution to ensure that the experiment is performed safely.

- ....

Method

1   Describe how you will carry out the experiment to show that the amplitude A decreases exponentially with the number of swings n. Try to give some detail that enables accurate measurements to be made; for example, how will you avoid parallax error, how will you choose the value of the independent variable when the dependent variable is measured?

....

....

....

....

....

....

....

2   Describe how you can increase the air resistance acting on the mass.

....

....

3   How can you manage this change to ensure that other factors such as the period do *not* change?

....

....

....

....

Results

Draw a table of results that can be used to record and process the data from this experiment. You do not have to fill in the table. Include the correct units in the column headings.

Analysis, conclusion and evaluation

a   Describe and explain how you can analyse the data to show the relationship between A and n. Your account should include plotting a graph and explaining how the graph shows that the relationship $A = A_0 e^{-\lambda n}$ is obeyed. Draw a sketch of the graph you expect to obtain.

> **TIP**
>
> Label the axes of your graph.

....

....

....

....

b Describe and explain how your graph can be used to find a value for λ.

....

....

....

....

c If you have the opportunity, and your teacher has checked your plan, follow your own, or a fellow student's, plan and take readings. Tabulate your data in your table and work out a value for λ.

Having performed the experiment, give some more details on difficulties and improvements to your initial plan.

....

....

....

....

Practical investigation 11.4: Simple harmonic oscillation of a mass on a spring

The period of oscillation of a mass on a spring depends on the mass. In this practical you will investigate this relationship and use the formula given by **simple harmonic motion** theory to determine the spring constant of the spring that you use.

YOU WILL NEED

Equipment:
- retort stand, boss and clamp • G-clamp or heavy weight • steel spring
- mass hangers with slotted masses, 100 g • stopwatch.

Safety considerations
- Make sure you have read the Safety advice at the beginning of this book and listen to any advice from your teacher before carrying out this investigation.
- Make sure that the retort stand cannot topple. This can be done by placing a heavy weight on the base of the retort stand or using a G-clamp to clamp the stand to the bench.

Method

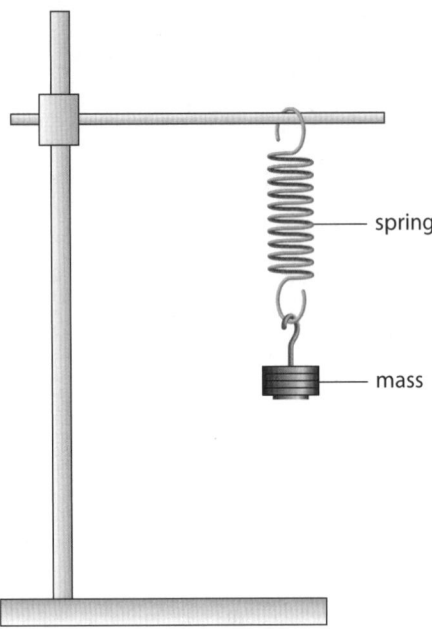

Figure 11.4: Spring suspended from rod with mass on other end.

1 Set up the apparatus as shown in Figure 11.4 so that vertical oscillations of the mass on the spring can be measured.

2 Start by placing about half the masses that you have available on the end of the spring and decide how to measure the time for one oscillation accurately (this is the period of the oscillation, T). You should think about how many oscillations, n, you can time, about where in the oscillation you should start and stop the timing, and whether you need a mark to indicate this place (this is called a fiducial mark).

3 Choose a range of masses to put on the end of the spring. The range should allow the period to vary noticeably, but you must make sure that the elastic limit of the spring is not exceeded. The mass on the end of the spring is M.

4 Measure the time for n oscillations of six different masses in the range you have chosen. Estimate the uncertainty in each measurement by taking three measurements for each value of t. Record all your readings in Table 11.3 in the Results section, and add your uncertainty after the $\pm$ for the average of t.

> **TIP**
>
> When the mass is removed the spring should not remain stretched!

> **TIP**
>
> The uncertainty is easiest found as half the difference between your maximum and minimum values of t.

Results

M / kg	n	Time for *n* oscillations *t* / s				T / s	lg (T / s)	lg (M / kg)
		1st	2nd	3rd	average			
						±	±	±	
						±	±	±	
						±	±	±	
						±	±	±	
						±	±	±	
						±	±	±	

Table 11.3: Results table.

Analysis, conclusion and evaluation

a   *T* is the period, or time for one oscillation. Complete Table 11.3 by calculating
    T / s, lg (*T* / s) and lg (*M* / kg) for each result. You can calculate the uncertainty for
    each value of lg (*T* / s) and assume that the uncertainty in *M* is very small and can
    be ignored.

b   Plot a graph on the next page of lg (*T* / s) on the *y*-axis and lg (*M* / kg) on the
    x-axis. Include error bars for lg *T*. Draw the straight line of best fit and a worst
    acceptable straight line on your graph.

c   Calculate the gradient of your best-fit line and use the worst-fit line to determine
    the uncertainty in the gradient. You do not need to give the unit.

> **TIP**
>
> Remember lg stands
> for lg to the base 10
> on your calculator.

    Gradient = ... ± ....

d   Theory suggests that $T = aM^b$, where *a* and *b* are constants.
    Write down lg *T* in terms of *M*, *a* and *b*.

    lg *T* = ...

e   Use your answers to parts **c** and **d** to determine the value of *b* and its uncertainty.
    You do not need to give the unit.

    b = ... ± ....

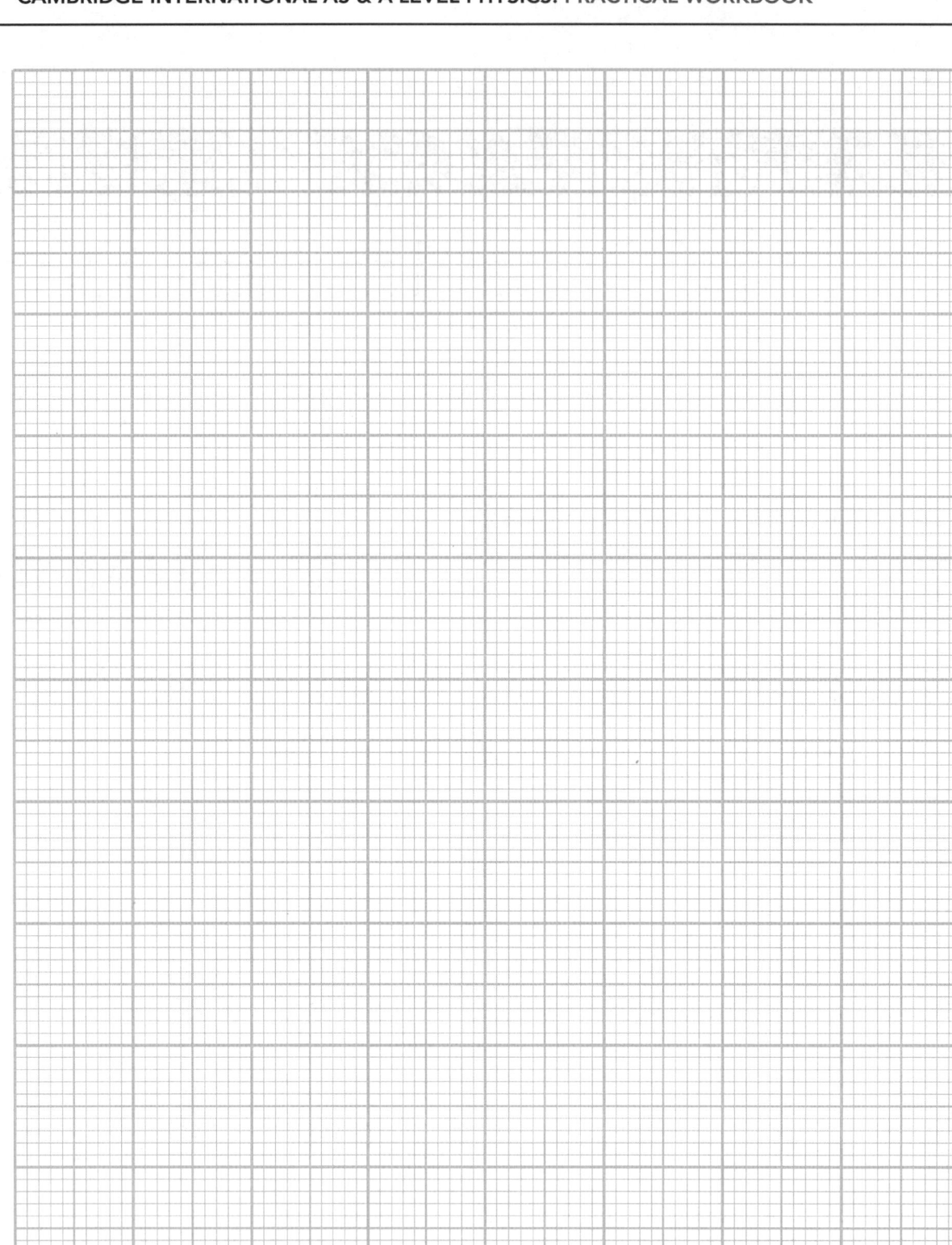

f Choose one point on your graph and use your value of b to calculate a value for a.

Point on graph: $\lg(T/s) = \dots\dots\dots\dots\dots$

$\lg(M/\text{kg}) = \dots\dots\dots\dots\dots$

$a = \dots\dots\dots\dots\dots$

g Simple harmonic motion theory suggests that $T = 2\pi\sqrt{\dfrac{M}{k}}$, where k is the spring constant.

Explain whether your answer for the constant b in part **e** agrees with this theory.

....

....

....

h Use your answer for the constant a in part **f** to find a value for k. You do not need to give a unit.

$k = \dots\dots\dots\dots\dots$

> **TIP**
>
> Combine the two equations for T to see how a depends on 2π and k.

> Chapter 12

Thermal physics and ideal gases

This chapter relates to Chapter 19: Thermal physics and Chapter 20: Ideal gases, in the coursebook. In this chapter you will complete investigations on:

- 12.1 Data analysis investigation into the thermocouple

- 12.2 Boyle's law

- 12.3 Planning investigation into specific latent heat of vaporisation of water

- 12.4 Data analysis investigation into specific latent heat of vaporisation of water.

Practical investigation 12.1: Data analysis The thermocouple

A student investigates the relationship between the e.m.f. produced by a **thermocouple** and the temperature difference between its two junctions. The voltage produced is low and so they use an operational amplifier with a gain of × 10.0. The apparatus is shown in Figure 12.1.

KEY WORD

thermocouple: a device consisting of wires of two different metals across which an e.m.f. is produced when the two junctions of the wires are at different temperatures

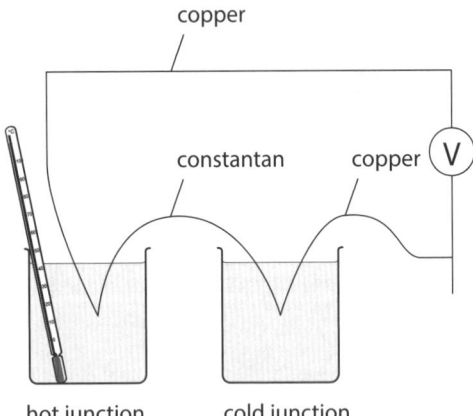

Figure 12.1: Circuit diagram with a hot junction and a cold junction.

The student measures the e.m.f. E of the thermocouple for different temperatures T of the hot junction of the thermocouple. The cold junction is kept in melting ice at 0 °C. The readings obtained are shown in Table 12.1.

Results

Temperature T / °C	E / mV	$\dfrac{E}{T}$ / mV °C^{-1}
10	0.390	0.0390 ± 0.0002
20	0.790	
30	1.20	
40	1.61	
50	2.04	
60	2.47	
70	2.91	
80	3.36	
90	3.81	
100	4.28	0.0428 ± 0.0002

Table 12.1: Results table.

Analysis, conclusion and evaluation

It is suggested that the e.m.f. E of the thermocouple varies with T according to the equation:

$E = aT + bT^2$

where a and b are constants.

a   Rearrange the formula to express $\dfrac{E}{T}$ in terms of a, b and T.

$\dfrac{E}{T} =$ ....

b   A graph is drawn with $\dfrac{E}{T}$ on the y-axis and T on the x-axis. What is the gradient and intercept of the graph in terms of a and b?

Gradient = ...   Intercept = ...

c   The uncertainties in the voltmeter reading produce an uncertainty of ± 5% in V.
     Complete Table 12.1 by calculating $\frac{E}{T}$ for all the readings. Add the uncertainty
     for each value of V. You may ignore any uncertainty in T.

d   Draw a graph of $\frac{E}{T}$ / mV °C^{-1} on the y-axis against T / °C on the x-axis.
     Include error bars for $\frac{E}{T}$. Draw a straight line of best fit and a straight line of
     worst fit on your graph.

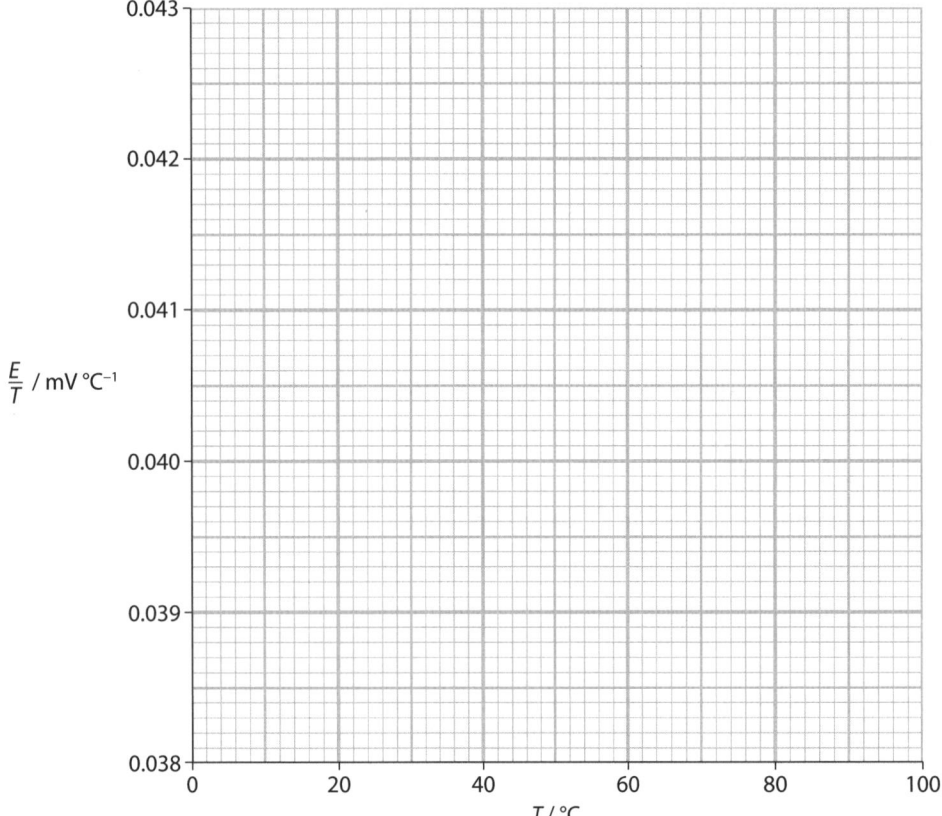

e   Calculate the gradient and intercept of your best-fit line and use the gradient and
     intercept of the worst-fit line to estimate the uncertainty in your values. Give units.

     Gradient of best-fit line = ... ± ....

     Intercept of best-fit line = ... ± ....

f   State the value of *a* and *b*, their uncertainties and their units.

$a = $ .... $\pm$ ....

$b = $ .... $\pm$ ....

g   A student uses the e.m.f. of a thermocouple thermometer to measure temperature. They also use a laboratory thermometer which involves the expansion of mercury.

Explain why, although both thermometers have been calibrated correctly, they may record different temperatures.

....

....

....

....

....

Practical investigation 12.2: Boyle's law

In this experiment you will investigate the relationship between the volume of a fixed mass of gas and its pressure.

YOU WILL NEED

Equipment:

• 10 ml disposable syringe, sealed at one end • stand and clamp to hold the syringe • 30 cm ruler • 100 g slotted mass hanger and nine 100 g slotted masses • loop of string.

Access to:

• a thermometer.

Safety considerations

•   Make sure you have read the Safety advice at the beginning of this book and listen to any advice from your teacher before carrying out this investigation.

- Make sure that the syringe is supported firmly upside down by the clamp and stand and that the loop of string is tied securely round the handle of the syringe. There should be about 15 cm from the bottom of the mass hanger to the bench to allow the handle to move down.

Method

1 Connect the apparatus securely as shown in Figure 12.2. Remove the mass hanger from the loop of string and check that there is about 6 ml of air in the syringe. Gently move the plunger up and down a few mm to ensure it does not stick.

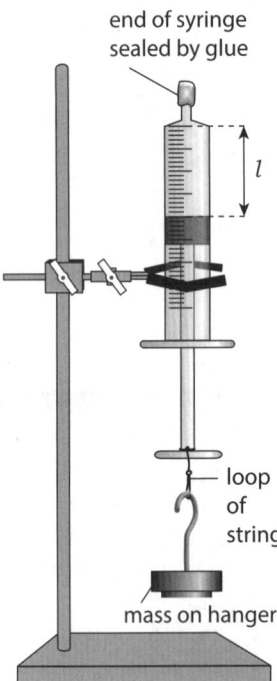

Figure 12.2: Mass attached to plunger of syringe.

2 Measure the length *l* of the air column from the plunger to the closed end of the syringe.

3 Hang a mass of 200 g from the loop of string and measure the new length *l*.

4 Hang further masses, increasing the mass by 200 g each time. Record the total mass *M* hanging from the syringe and the corresponding value of *l* in Table 12.2. Each time, move the plunger up and down so that it does not stick and repeat each reading at least once. Give your value for *l* in metres; for example, 4.0 cm = 0.040 m.

5 Record room temperature.

Results

M / kg	l / m			$\frac{1}{l}$ / m^{-1}
	1st reading	2nd reading	average and uncertainty	
0			±	±
0.200			±	±
0.400			±	±
0.600			±	±
0.800			±	±
1.000			±	±

Table 12.2: Results table.

Room temperature = ...

Analysis, conclusion and evaluation

The **gas law equation** $pV = nRT$ applied to the air in the syringe produces the expression:

$$(p_0 A - Mg)l = nRT$$

or $\dfrac{1}{l} = \dfrac{p_0 A}{nRT} - \dfrac{Mg}{nRT}$

where p_0 is atmospheric pressure, A is the cross-sectional area of the syringe, T is the temperature of the gas in K and the other symbols have their usual meaning for an ideal gas.

a   You will plot a graph of $\dfrac{1}{l}$ on the y-axis against M on the x-axis.

Using the symbols in the gas law equation as applied to the air in the syringe, determine expressions for the gradient and intercept of the graph.

Gradient = ...   Intercept = ...

b   Complete Table 12.2 by calculating values for $\dfrac{1}{l}$ / m^{-1}. Include values for uncertainty.

c   Plot a graph of $\dfrac{1}{l}$ / m^{-1} on the y-axis against M / kg on the x-axis using the grid on the next page. Include error bars for $\dfrac{1}{l}$.

Draw the straight line of best fit and a worst acceptable straight line on your graph.

KEY EQUATION

ideal gas equation

$$pV = nRT$$

TIP

If both your readings are the same the uncertainty is not zero, it is the smallest division on the ruler that you are using.

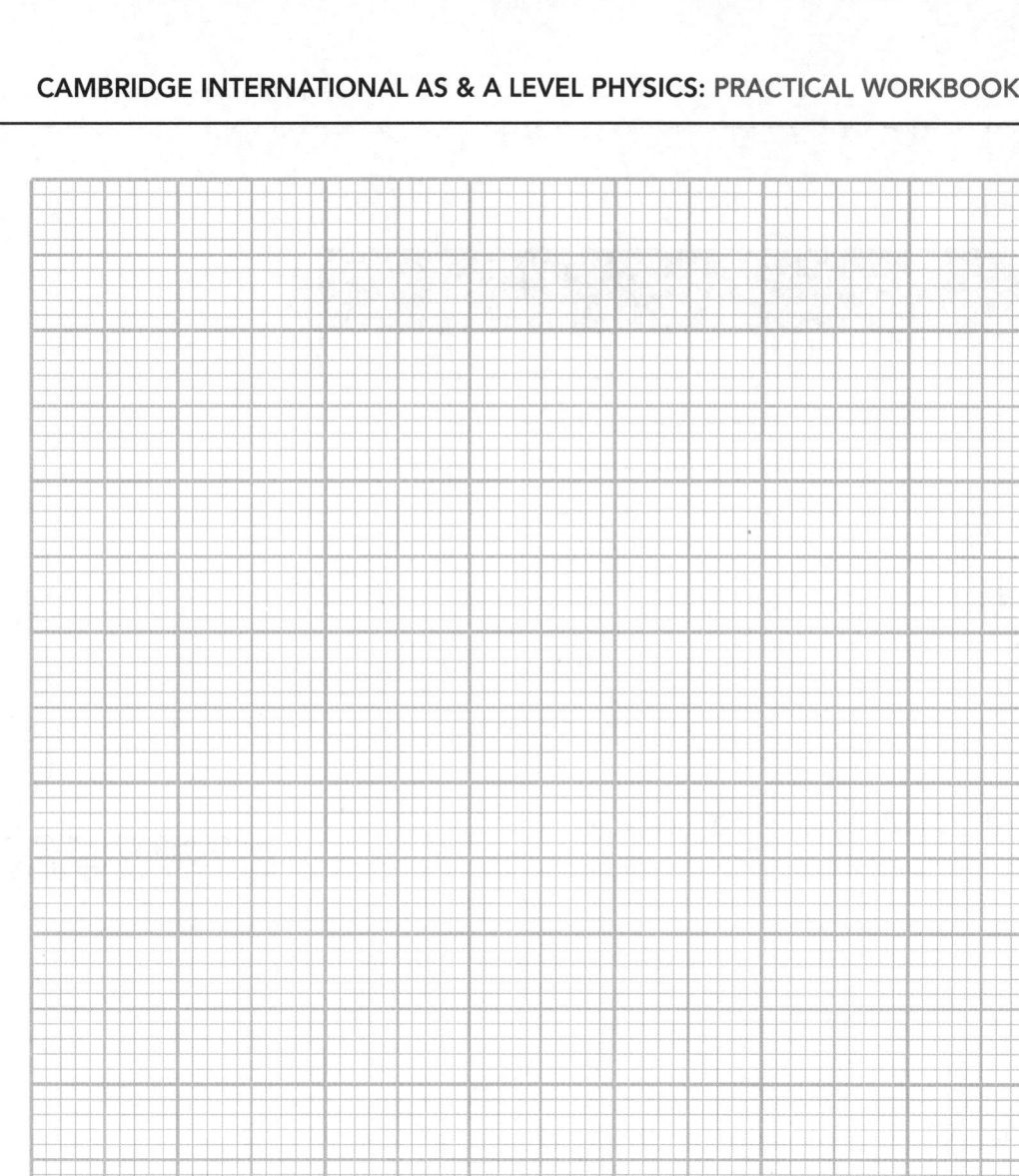

d   Determine the gradient of the line of best fit. Include the absolute uncertainty in your answer by using the gradient of the line of worst fit.

Gradient = ... ± ....

e   Convert your value of room temperature into kelvin (K).

Room temperature = ... K

f   Use your answers to parts **a**, **d** and **e** to find the number of moles of gas in the syringe. The value of the gas constant R is 8.31 J K^{-1} mol^{-1}. Include the uncertainty in your answer.

Number of moles n = ... ± ....

g   Explain how atmospheric pressure can be obtained using the intercept of your graph, the area A, n, R and T.

....

....

h   State and explain how the gradient and intercept of your graph would differ if the initial volume of air trapped in the syringe was halved.

....

....

Practical investigation 12.3: Planning Specific latent heat of vaporisation of water

A student is investigating how the power of a heater affects the amount of water that is lost during boiling. They set up the apparatus as shown in Figure 12.3, and attach some electrical meters. The power of the heater is sufficient, when connected to a power supply of 15 V, to boil water at a reasonable rate.

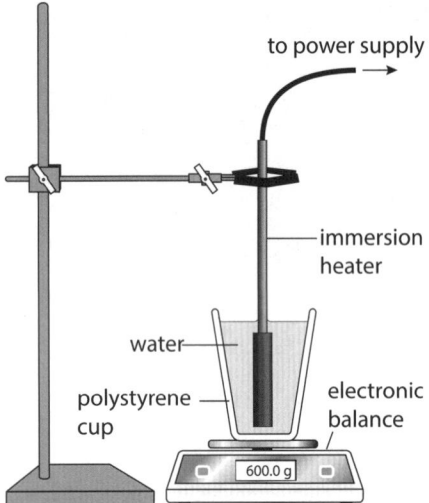

Figure 12.3: Immersion heater in polystyrene cup filled with water on balance.

It is suggested that, assuming the water is at its boiling point, the relationship between the power P of the heater and the mass m of water that turns to steam in a time t is given by the equation:

$$Pt = mL + H$$

where L is the **specific latent heat of vaporisation** of water and H is the thermal energy lost from the apparatus during this time.

The amount of thermal energy H lost from the apparatus depends on a number of factors, in particular the length of time for which the heater is switched on.

You are going to design a laboratory experiment to test the relationship between P and m, explaining how the latent heat of vaporisation is determined by a graphical method.

In your account you should:

- list the dependent variable, independent variable and variables to be controlled

- draw a circuit diagram showing how the meters are connected

- write an account of the procedure to be followed, including the measurements made and how the quantities are calculated. Give as many details as possible to ensure that the experiment is reliable and that heat loss is reduced and is as constant as possible

- state the safety precautions to be taken

- draw a sketch diagram of the graph the student should plot

- explain how the graph can be used to obtain the latent heat of vaporisation.

Variables

List the dependent variable, the independent variable and the variables that should be controlled. The variables to be controlled are quantities that must be kept the same.

> **KEY WORDS**
>
> **specific latent heat of vaporisation:**
> the amount of heat energy per unit mass needed to convert unit mass of solid to liquid without change in temperature

> **TIP**
>
> The time t must be kept constant or the heat loss H will not be a constant.

- Dependent variable: ....

- Independent variable: ....

- Variables to be controlled: ....

....

YOU WILL NEED

List the extra equipment you will need, apart from that shown in Figure 12.3, and draw a labelled diagram of how you will set up the apparatus in order to obtain the necessary measurements.

- ....

- ....

- ....

- ....

- ....

- ....

- ....

- ....

Circuit diagram

Safety considerations

- Make sure you have read the Safety advice at the beginning of this book and listen to any advice from your teacher before carrying out this investigation.

- State one safety precaution that is relevant to this experiment.

....

....

Method

Describe how you will carry out the experiment.

....

....

....

....

....

....

> **TIP**
>
> Try to give some detail that explains how you will try to ensure measurements are accurate.
> For example, how can you ensure the current is constant, how can you reduce heat loss or how can you reduce the percentage uncertainty in time?

Results

Draw a table·of results which can be used to record and process the data from this experiment. You do not have to fill in the table. Remember to include the correct units in the column headings.

Analysis, conclusion and evaluation

a   Describe how you can analyse the data to show the relationship between P and m. Your account should include plotting a graph and using either the gradient or intercept of the graph. You could draw a sketch of the graph you expect if the equation is valid.

....

....

....

....

....

....

b   Suggest how a similar experiment could be performed to measure the latent heat of fusion of water and suggest how the graph would differ from that for water being boiled.

....

....

....

....

....

....

Practical investigation 12.4:
Data analysis
Specific latent heat of vaporisation of water

The investigation described in the introduction to Practical investigation 12.3 is carried out.

The student switches on the power supply. Once the water is boiling steadily, they record the mass m of water boiled away in 3.0 minutes. This reading is then repeated. While heating, the student records the potential difference across and the current in the heater several times to obtain average values. They alter the supply voltage and repeat the experiment. There was very little fluctuation in the current and voltage readings and uncertainties in these quantities can be ignored.

Table 12.3 shows the results.

Results

Voltage / V	Current / A	Power / W	m / g			
			1st	2nd	average	uncertainty
6.4	5.1		1.9	2.5		
7.5	6.1		2.6	3.1		
8.4	6.9		3.8	4.2		
8.6	6.9		3.9	4.2		
8.9	7.2		4.5	4.9		

Table 12.3: Results table.

Analysis, conclusion and evaluation

a To verify the relationship $Pt = mL + H$, a graph is plotted with m on the y-axis and P on the x-axis. Rearrange the equation to make m the subject of the formula and write down expressions for the gradient and intercept of this graph, in terms of t, L and H.

m = ... Gradient = ... Intercept = ...

b Calculate and record values of P / W and the average value of m / g in Table 12.3. Calculate the absolute uncertainties for m.

c Plot a graph of m / g on the y-axis against P / W on the x-axis. Include error bars for m / g.

TIP
power = VI

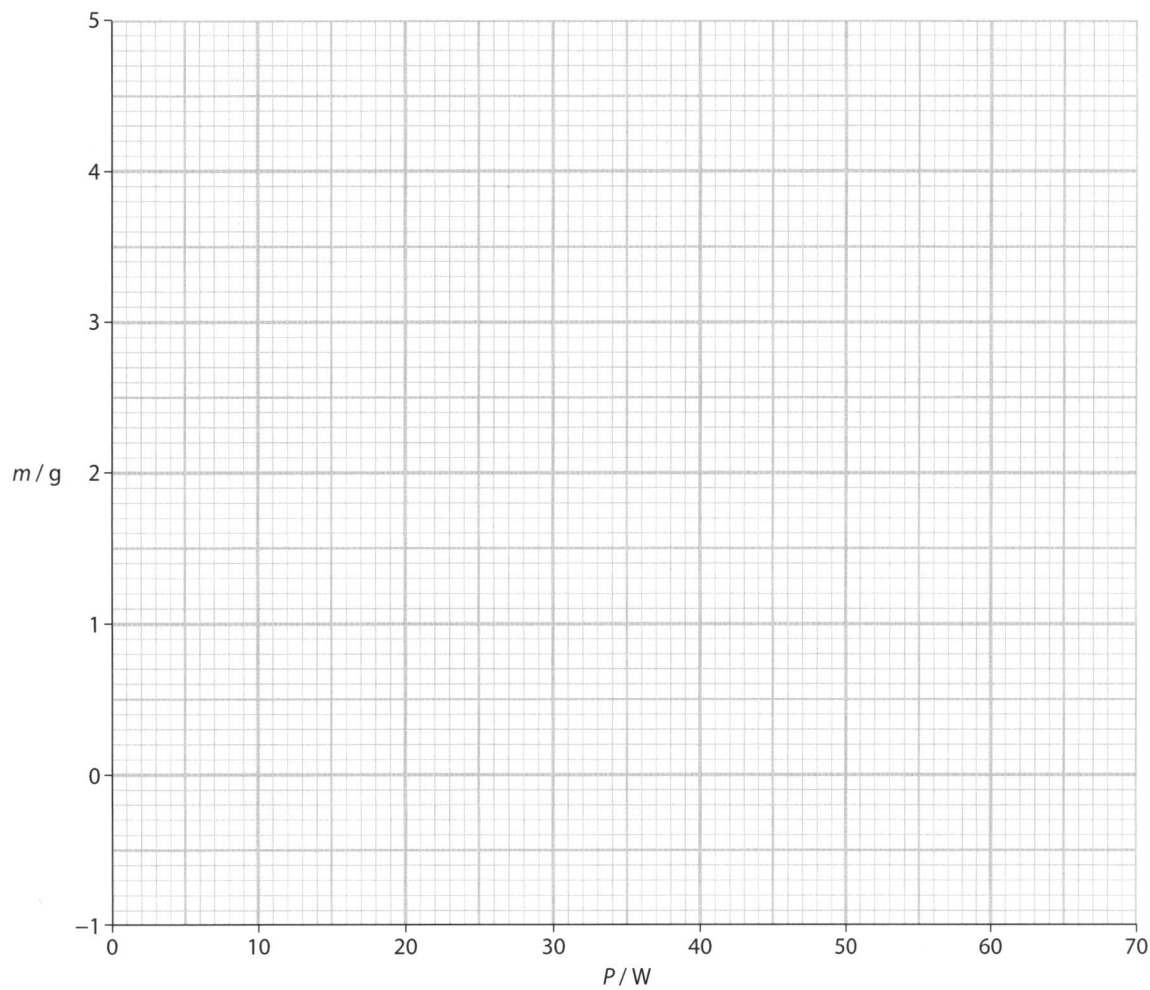

d Draw the straight line of best fit and a worst acceptable straight line on your graph.

e Determine the gradient of the line of best fit. Include the absolute uncertainty in your answer by using the gradient of the worst-fit line.

Gradient = ... ± ....

f Determine the intercept of your graph. Include the unit of your answer.

Intercept = ...

TIP

The intercept is the value of m on your graph when $P = 0$.

g Using your answers to parts **a**, **e** and **f**, determine the values of L and H. Include appropriate units. The time t of heating is 180 s.

$L =$ .... $H =$ ....

h The uncertainty in measuring t is ± 2 s. Use this fact and the uncertainty in the gradient to determine the percentage uncertainty in L.

Percentage uncertainty in $L =$ ....%

i Explain why the water must be boiling steadily before measuring m.

....

....

....

j Describe how the graph would differ if the heating was for 6 minutes instead of 3 minutes.

....

....

....

> Chapter 13

Coulomb's law and capacitance

CHAPTER OUTLINE

This chapter relates to Chapter 22: Coulomb's law and Chapter 23: Capacitance, in the coursebook.
In this chapter you will complete investigations on:

- 13.1 Planning investigation into how the time for the potential difference across a capacitor to halve varies with the resistance

- 13.2 Determination of the capacitance of a capacitor in a d.c. circuit

- 13.3 Planning investigation into how the peak current in a capacitor circuit varies with the frequency of the a.c. supply

- 13.4 Determination of the capacitance of a capacitor in an a.c. circuit

- 13.5 Planning investigation into how the resistance of a thermistor varies with temperature.

Practical investigation 13.1: Planning
How the time for the potential difference across a capacitor to halve varies with the resistance

When a capacitor is discharged through a resistor, as shown in Figure 13.1, the potential difference (p.d.) across the capacitor reduces.

Figure 13.1: Circuit diagram.

Theory suggests that the relationship between the **capacitance** C of the capacitor, the resistance R of the resistor and the time t for the p.d. across the capacitor to halve is given by:

$$\ln\left(\frac{1}{2}\right) = -\frac{t}{CR}$$

KEY WORDS

capacitance (of a capacitor): the charge stored on one plate per unit potential difference between the plates

You are going to design a laboratory experiment to test the relationship between t and R. In your account you will:

- write an account of the procedure to be followed
- describe the measurements to be taken
- describe the types of variables involved
- describe how the data is analysed to determine a value for C
- give one or two safety precautions that should be taken.

Variables

List the dependent variable, the independent variable and the variables that should be controlled. The variables to be controlled are quantities that must be kept the same.

- Dependent variable: ....

- Independent variable: ....

- Variables to be controlled: ....

    ....

YOU WILL NEED

List the equipment you will need and draw a labelled diagram of how you will set up the apparatus to obtain the necessary measurements.

- ....

- ....

- ....

- ....

- ....

- ....

- ....

Safety considerations

- Make sure you have read the Safety advice at the beginning of this book and listen to any advice from your teacher before carrying out this investigation.

- ....

- ....

Method

Describe how you will carry out the experiment.

....

....

....

....

....

....

....

....

....

Results

Draw a table of results which can be used to record and process the data from this experiment. You do not have to fill in the table. Remember to include the correct units in the column headings.

Data analysis

a   Describe how you can analyse the data to show the relationship between t and R. Your account should include plotting a graph and using the gradient of the graph to determine a value for C.

....

....

....

....

Practical investigation 13.2: Determination of the capacitance of a capacitor in a d.c. circuit

A capacitor is charged by applying an e.m.f. across it. It is then discharged through a resistor.

YOU WILL NEED
Equipment:
• capacitor • 100 kΩ resistor • ammeter (0–100 μA) • power supply • switch • connecting leads.

Safety considerations

*   Make sure you have read the Safety advice at the beginning of this book and listen to any advice from your teacher before carrying out this investigation.

*   A low voltage power supply should be used.

*   Ensure that the polarity of the capacitor is correct.

Method

1   Set up the circuit shown in Figure 13.2. You will need to make sure that the capacitor is connected correctly.

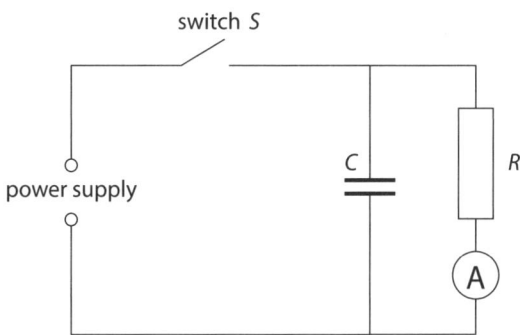

Figure 13.2: Circuit diagram.

2   Close switch *S*.

3   Measure and record the current. This is the current when *t* = 0.

4   Open switch *S*.

5   Measure and record the current *I* every 10 s in Table 13.1 in the Results section.

6   Include the absolute uncertainty in your readings of current.

TIP

Check the polarity of the ammeter and the capacitor.

Results

R = ...

t / s	I / μA	ln (I) / μA
	±	±
	±	±
	±	±
	±	±
	±	±
	±	±
	±	±
	±	±

Table 13.1: Results table.

Analysis, conclusion and evaluation

a   Calculate the value of ln (*I* / μA) in Table 13.1. Include the uncertainties in ln *I*.

b   Draw a graph of ln (*I*) / μA on the *y*-axis plotted against *t* / s on the *x*-axis.

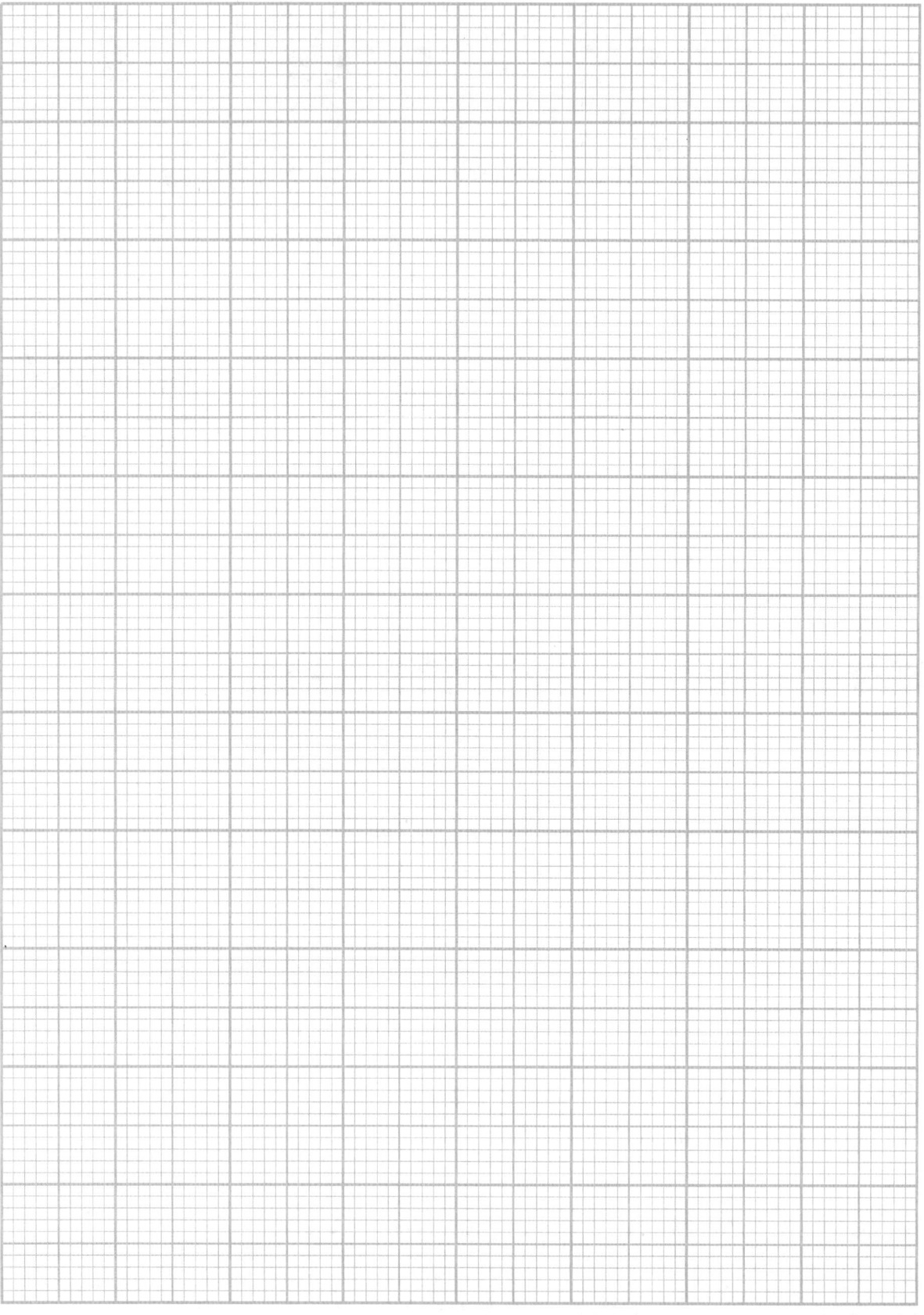

c It is suggested that the relationship between I and t is:

$$I = I_0 e^{-\frac{t}{CR}}$$

where C is the capacitance of the capacitor and R is the resistance of the resistor.
By rearranging the relationship, make ln I the subject of the formula.

$I = $ ....

d Using your equation from part **c**, determine the gradient and y-intercept of your graph of ln I against t in terms of C, R and I.

Gradient = ... y-intercept = ...

e Use the uncertainty in your values of ln I to draw error bars on your graph.
On your graph, draw a straight line of best fit and a worst acceptable straight line.

f Determine the gradient of the line of best fit and the gradient of the line of worst fit. You do not need to give units. Estimate the uncertainty in your value of the gradient.

Gradient of best-fit line = ...

Gradient of worst-fit line = ...

Uncertainty in gradient = ...

g Determine the y-intercept of the line of best fit and the y-intercept of the line of worst fit. You do not need to give units. Estimate the uncertainty in your value of the y-intercept.

y-intercept of best-fit line = ...

y-intercept of worst-fit line = ...

Uncertainty in y-intercept = ...

h Using your values of the gradient and y-intercept and your value for R, determine the values for C and I_0.

$C = $ .... $I_0 = $ ....

i Determine the percentage uncertainty for C and I_0.

Percentage uncertainty in $C = $ ....%

Percentage uncertainty in $I_0 = $ ....%

j Explain any inaccuracies you had in determining C.

....

....

....

....

> **TIP**
>
> Consider the quantities used to plot the graph and determine C.

k Explain the impact on your values of C and I_0.

....

....

....

....

....

Practical investigation 13.3: Planning How the peak current in a capacitor circuit varies with the frequency of the a.c. supply

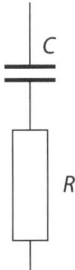

Figure 13.3: Capacitor and resistor in series.

Theory suggests that the relationship between the peak current I_0 and the frequency f of the alternating current is given by:

$$I_0 = kCV_0 f^n$$

where C is the capacitance of the capacitor, V_0 is the peak potential difference applied to the circuit and k and n are constants.

You are going to design a laboratory experiment to test the relationship between I_0 and f. In your account you will:

- write an account of the procedure to be followed

- describe the measurements to be taken

- describe the types of variables involved

- describe how the data is analysed to determine a value for k and n

- give one or two safety precautions that may be taken.

Variables

List the dependent variable, the independent variable and the variables that should be controlled. The variables to be controlled are quantities that must be kept the same.

- Dependent variable: ....

- Independent variable: ....

- Variables to be controlled: ....

  ....

YOU WILL NEED

List the equipment you will need and draw a labelled diagram of how you will set up the apparatus to obtain the necessary measurements.

- ....

- ....

- ....

- ....

- ....

Draw a circuit diagram of how to set up the apparatus.

Safety considerations

- ....

Method

Describe how you will carry out the experiment.

....

....

....

....

....

....

....

Results

Draw a table of results which can be used to record and process the data from this experiment. You do not have to fill in the table. Remember to include the correct units in the column headings.

Analysis, conclusion and evaluation

a   Describe how you can analyse the data to show the relationship between I_0 and f. Your account should include plotting a graph and using the gradient and y-intercept of the graph to determine a value for k and n.

....

....

....

....

....

....

....

....

Practical investigation 13.4: Determination of the capacitance of a capacitor in an a.c. circuit

A capacitor is connected to a signal generator. The peak current is recorded for different frequencies.

YOU WILL NEED

Equipment:

- capacitor • 100 Ω resistor • cathode ray oscilloscope • signal generator
- connecting leads.

Safety considerations

- Make sure you have read the Safety advice at the beginning of this book and listen to any advice from your teacher before carrying out this investigation.

- There are no specific safety considerations for this practical.

Method

1   Set up the circuit shown in Figure 13.4. Do not change the output of the signal generator.

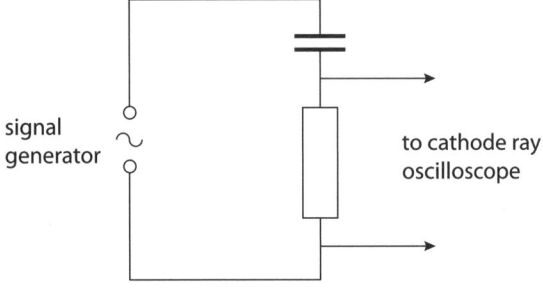

Figure 13.4: Circuit diagram.

2   Measure and record the frequency f from the signal generator in Table 13.2 in the Results section.

3   Measure and record the peak voltage V_{max} reading from the oscilloscope in Table 13.2.

4   Repeat steps **2** and **3** until you have six different values of frequency.

5   Measure the peak voltage V_0 from the signal generator.

6   Record the value of R.

Results

$V_0 =$ .... $R =$ ....

f / Hz		V_{max} / V		
		±	±	±
		±	±	±
		±	±	±
		±	±	±
		±	±	±
		±	±	±

Table 13.2: Results table.

TIP

Remember to label all your columns in the table of results with a quantity and unit.

Analysis, conclusion and evaluation

a   Calculate the values of $\dfrac{1}{f}$ / Hz^{-1} and enter them in Table 13.2.

b   Calculate the values of the peak current I_0 and the values of X / Ω and add them to Table 13.2. Use the following equations and include the uncertainties in I_0 and X:

$$I_0 = \frac{V_{max}}{R} \text{ and } X = \frac{V_0}{I_0}$$

c   Draw a graph of X / Ω on the y-axis plotted against $\dfrac{1}{f}$ / Hz^{-1} on the x-axis.

d   It is suggested that the relationship between X and $\dfrac{1}{f}$ is:

$$X = \frac{1}{2\pi fC}$$

where C is the capacitance of the capacitor.

Use this equation to determine the gradient of your graph of X against $\dfrac{1}{f}$.

Gradient = ...

e   Use the uncertainty in your values of X to draw error bars on your graph. On your graph, draw a straight line of best fit and a worst acceptable straight line.

f   Determine the gradient of the line of best fit and the gradient of the line of worst fit. You do not need to give units. Estimate the uncertainty in your value of the gradient.

Gradient of best-fit line = ...

Gradient of worst-fit line = ...

Uncertainty in gradient = ...

g   Using your value of the gradient, determine a value for C. Include appropriate units for C.

$C =$ ....

h   Determine the percentage uncertainty in C.

Percentage uncertainty in $C =$ ....%

i   Determine the absolute uncertainty in C.

Absolute uncertainty in $C =$ ....

TIP

Make sure you use the correct combination of maximum and minimum values to determine the absolute uncertainty.

j   Explain how the frequency of the a.c. could be determined accurately.

....

....

k   Explain how V_0 can be measured accurately.

....

....

Practical investigation 13.5: Planning How the resistance of a thermistor varies with temperature

KEY WORD

thermistor: a device whose electrical resistance changes when its temperature changes

The resistance of a **thermistor** depends on the temperature of the thermistor.

Theory suggests that the relationship between the resistance R of a thermistor and its absolute temperature T is:

$R = pT^q$

where p and q are constants.

You are going to design a laboratory experiment to test the relationship between R and T. In your account you will:

- write an account of the procedure to be followed

- describe the measurements to be taken

- describe the types of variables involved

- describe how the data is analysed to determine p and q

- give one or two safety precautions that may be taken.

Variables

List the dependent variable, the independent variable and the variables that should be controlled. The variables to be controlled are quantities that must be kept the same.

- Dependent variable: ....

- Independent variable: ....

- Variables to be controlled: ....

  ....

YOU WILL NEED

List the equipment you will need and draw a labelled diagram of how you will set up the apparatus to obtain the necessary measurements.

- ....
- ....
- ....
- ....
- ....
- ....
- ....
- ....

Safety considerations

- Make sure you have read the Safety advice at the beginning of this book and listen to any advice from your teacher before carrying out this investigation.

- ....

- ....

Method

Describe how you will carry out the experiment.

....

....

....

....

....

Results

Draw a table of results which can be used to record and process the data from this experiment. You do not have to fill in the table. Remember to include the correct units in the column headings.

Analysis, conclusion and evaluation

a   Describe how you can analyse the data to show the relationship between R and T. Your account should include plotting a graph and using the gradient and intercept of the graph to determine the constants p and q.

....

....

....

....

....

....

⟩ Chapter 14

Magnetic fields, electromagnetism and charged particles

CHAPTER OUTLINE

This chapter relates to Chapter 24: Magnetic fields and electromagnetism and Chapter 25: Motion of charged particles, in the coursebook.

In this chapter you will complete investigations on:

- 14.1 The variation of the force on a conductor in a magnetic field

- 14.2 Planning investigation into how the separation of two foils carrying a current varies with the current

- 14.3 Planning investigation into the magnetic field of a coil using a Hall probe

- 14.4 How the strength of a magnetic field in a coil varies

- 14.5 Observing charged particles.

Practical investigation 14.1: The variation of the force on a conductor in a magnetic field

A magnet is placed on a balance. A current is passed through several turns of a wire which are placed between the poles of the magnet resulting in a force being exerted on the scales. The force is determined from the change in the balance reading.

KEY WORDS

magnetic field: a force field in which a magnet, a wire carrying a current, or a moving charge experiences a force

YOU WILL NEED

Equipment:

- copper wire • pair of magnets and a yoke • ammeter • two retort stands and clamps • connecting leads • mass balance • high current power supply.

Safety considerations

- Make sure you have read the Safety advice at the beginning of this book and listen to any advice from your teacher before carrying out this investigation.

Method

1 Set up the apparatus as shown in Figure 14.1.

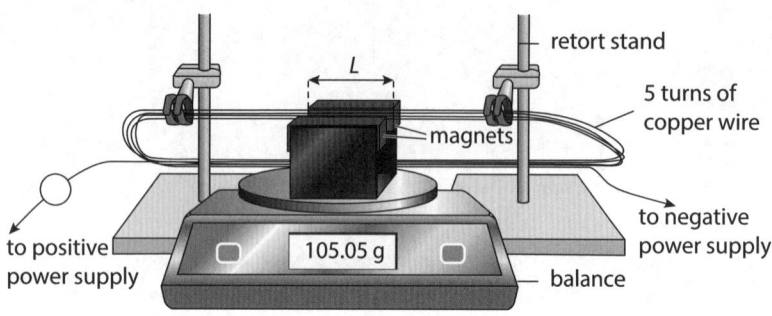

Figure 14.1: Copper wire between magnets.

The two magnets have their poles on their sides and not at their ends. Place them on the yoke so that opposite poles are facing each other across the space between them. Check with your teacher if you do not know how to place the magnets on the yoke.

2 Measure and record the length L of the magnet in the Results section. Include the absolute uncertainty in L. Measure and record the number of turns N of wire in the Results section.

3 Measure and record the balance reading R_0 without any current in the Results section. Include the absolute uncertainty in R_0.

4 Switch on the power supply. Measure and record the current I and the balance reading R. Switch off the power supply. Record your results in Table 14.1 in the Results section. Include the absolute uncertainty in R.

5 Repeat step **4** until you have six different values of current.

Results

$L =$ .... $R_0 =$ ....

$N =$ ....

I / A	R / g		(R − R₀) / g	
		±		±
		±		±
		±		±
		±		±
		±		±
		±		±

Table 14.1: Results table.

Analysis, conclusion and evaluation

a Calculate the values of $(R - R_0)\,/\,g$ and add them to Table 14.1. Include the uncertainties in $(R - R_0)$.

TIP

Remember the rule for combining uncertainties when subtracting quantities.

b Draw a graph with $(R - R_0)\,/\,g$ on the y-axis plotted against $I\,/\,A$ on the x-axis. It is suggested that the relationship between $(R - R_0)$ and I is

$(R - R_0)g = NBIL$

where g is the acceleration of free fall, B is the **magnetic flux density**, L is the length of the magnet and N is the number of turns of wire.

c Using the equation, determine the gradient of your graph of $(R - R_0)$ against I in terms of B.

KEY WORDS

magnetic flux density: the force acting per unit current per unit length on a wire placed at right angles to the magnetic field

Gradient = ...

d Use the uncertainty in your values of $(R - R_0)$ to draw error bars on your graph. Draw a straight line of best fit and a worst acceptable straight line.

e Determine the gradient of the line of best fit and the gradient of the line of worst fit. You do not need to give units. Estimate the uncertainty in your value of the gradient.

Gradient of best-fit line = ...

Gradient of worst-fit line = ...

Uncertainty in gradient = ...

f Using your value of the gradient, determine a value for B. Include appropriate units for B.

B = ...

g Determine the percentage uncertainty for B.

Percentage uncertainty in B = ...%

h Explain how you arranged the apparatus so that the readings for $(R - R_0)$ were a maximum.

....

....

....

....

i   Explain how the readings for $(R - R_0)$ could be increased.

....

....

Practical investigation 14.2: Planning How the separation of two foils carrying a current varies with the current

When current I flows through two parallel foils, as shown in Figure 14.2, the foils separate a distance s. The dashed lines show the position of the foils when the current is zero.

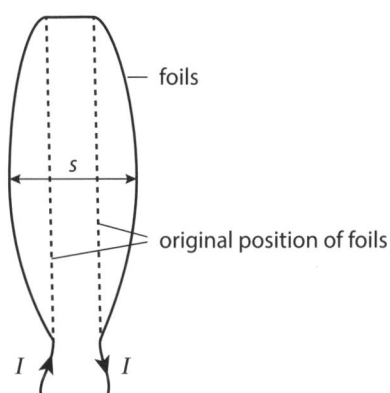

Figure 14.2: Current flowing through two parallel foils.

It is suggested that the relationship between the separation s and the length L of each foil is given by:

$$s = kL^m$$

where k and m are constants.

You are going to design a laboratory experiment to test the relationship between s and L. In your account you will:

*   write an account of the procedure to be followed

*   describe the measurements to be taken

*   describe the types of variables involved

*   describe how the data can be analysed to determine values for k and m

*   give one or two safety precautions that should be taken.

Variables

List the dependent variable, the independent variable and the variables that should be controlled. The variables to be controlled are quantities that must be kept the same.

- Dependent variable: ....

- Independent variable: ....

- Variables to be controlled: ....

  ....

YOU WILL NEED

List the equipment you will need and draw a labelled diagram of how you will set up the apparatus to obtain the necessary measurements.

- ....
- ....
- ....
- ....
- ....
- ....
- ....
- ....
- ....

Safety considerations

- Make sure you have read the Safety advice at the beginning of this book and listen to any advice from your teacher before carrying out this investigation.

....

....

Method

Describe how you will carry out the experiment.

....

....

....

....

....

....

....

....

....

> **TIP**
>
> Remember to include detail on how to measure s accurately.

Results

Draw a table of results that can be used to record and process the data from this experiment. You do not have to fill in the table. Remember to include the correct units in the column headings.

Analysis, conclusion and evaluation

a   Describe how you can analyse the data to show the relationship between s and L. Your account should include plotting a graph and using the gradient and y-intercept of the graph to determine values for k and m.

....

....

....

....

Practical investigation 14.3: Planning The magnetic field of a coil using a Hall probe

A Hall probe may be used to measure the strength of a magnetic field. The probe has a thin slice of semiconductor material. A small potential difference (called the **Hall voltage**) is created across the width of the semiconductor; the magnitude of the potential difference is directly proportional to the strength of the magnetic field. Figure 14.3 shows a typical probe.

Figure 14.3: Hall probe.

A student is investigating how the strength of the magnetic field B at a distance d from the centre of a current-carrying solenoid varies with the distance d, as shown in Figure 14.4.

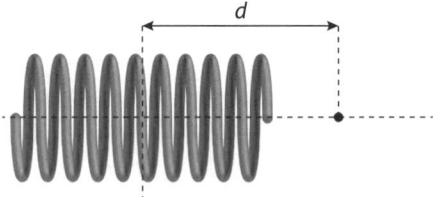

Figure 14.4: Solenoid.

> ### KEY WORDS
>
> **Hall voltage:** the potential difference produced across the sides of a conductor when an external magnetic field is applied perpendicular to the direction of the current; the Hall voltage V_H is directly proportional to the magnetic flux density B

The student suggests that:

$B = B_0 e^{-kd}$

where B_0 is the strength of the magnetic field at the centre of the coil and k is a constant.

You are going to design a laboratory experiment based on Figure 14.4 to test the relationship between B and d. In your account you will:

- write an account of the procedure to be followed

- describe the measurements to be taken

- describe the types of variables involved

- describe how the data is analysed

- give one or two safety precautions that may be taken.

Variables

List the dependent variable, the independent variable and the variables that should be controlled. The variables to be controlled are quantities that must be kept the same.

- Dependent variable: ....

- Independent variable: ....

- Variables to be controlled: ....

  ....

YOU WILL NEED

List the equipment you will need and draw a labelled diagram of how you will set up the apparatus in order to obtain the necessary measurements.

- ....

- ....

- ....

- ....

- ....

- ....

- ....

Safety considerations

- Make sure you have read the Safety advice at the beginning of this book and listen to any advice from your teacher before carrying out this investigation.

- ....

....

Method

Describe how you will carry out the experiment.

....

....

....

....

....

....

....

....

....

....

> **TIP**
>
> Describe how d could be measured accurately.

Results

Draw a table of results which can be used to record and process the data from this experiment. You do not have to fill in the table. Remember to include the correct units in the column headings.

Analysis, conclusion and evaluation

a   Describe how you can analyse the data to test the relationship between B and d. Your account should include plotting a graph and using the gradient and y-intercept of the graph to determine values for B_0 and k.

....

....

....

....

....

....

....

Practical investigation 14.4: How the strength of a magnetic field in a coil varies

A Hall probe has a thin slice of semiconductor material carrying a current. When the slice is at right angles to a magnetic field a (Hall) voltage is induced across the slice. This voltage is proportional to the strength of the magnetic field.

YOU WILL NEED

Equipment:
- long metal coil • variable resistor • ammeter • power supply • Hall probe
- voltmeter • metre rule • connecting leads.

Safety considerations
- Make sure you have read the Safety advice at the beginning of this book and listen to any advice from your teacher before carrying out this investigation.

- The metallic coils may become hot.

Method

1   Set up the coil as shown in Figure 14.5. You will need to make sure that the Hall probe is connected correctly.

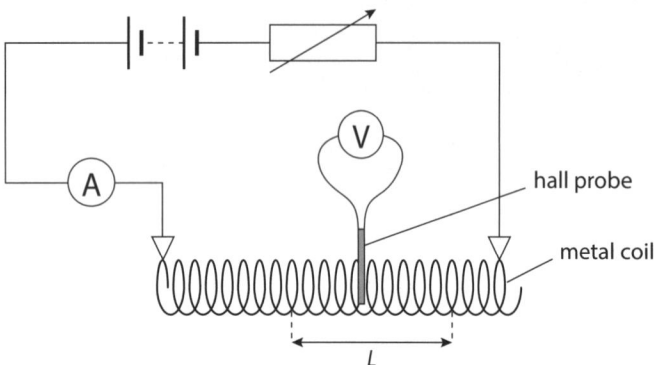

Figure 14.5: Circuit diagram.

2   Measure and record the length L of ten of the coils in Table 14.2 in the Results section. Include the absolute uncertainty in L.

3   Place the Hall probe in the centre of the coil. Ensure that the slice of the probe is at right angles to the axis of the coil. Measure and record the (Hall) voltage V in Table 14.2 in the Results section.

4   You just need to stretch the coil to achieve a different value for L. You do not need to change the connections to the coil. Measure and record the Hall voltage V for different lengths L in Table 14.2.

5   Measure and record the current I in the Results section. Include the absolute uncertainty in I.

Results

$N = 10$ $I = $ .... $\pm$ ....

$\pm$	$\pm$	
$\pm$	$\pm$	
$\pm$	$\pm$	
$\pm$	$\pm$	
$\pm$	$\pm$	
$\pm$	$\pm$	

Table 14.2: Results table.

Analysis, conclusion and evaluation

a Calculate the values of $\dfrac{1}{L}$ / m^{-1} and add them to Table 14.2. Include the uncertainties in $\dfrac{1}{L}$.

b Draw a graph with V / V on the y-axis plotted against $\dfrac{1}{L}$ / m^{-1} on the x-axis.

c It is suggested that the relationship between V and $\dfrac{1}{L}$ / m^{-1} is:

$$V = \frac{kNI}{L}$$

where k is a constant.

Using this relationship, determine the gradient of your graph of V / V against $\dfrac{1}{L}$ in terms of k, N and I.

Gradient = ...

d Use the uncertainty in your values of $\dfrac{1}{L}$ to draw error bars on your graph.

Draw a straight line of best fit and a worst acceptable straight line.

e Determine the gradient of the line of best fit and the gradient of the line of worst fit. You do not need to give units. Estimate the uncertainty in your value of the gradient.

Gradient of best-fit line = ...

Gradient of worst-fit line = ...

Uncertainty in gradient = ...

f Using your value of the gradient, determine a value for k. Include an appropriate unit for k.

k = ...

g Determine the percentage uncertainty for k.

Percentage uncertainty in k = ...%

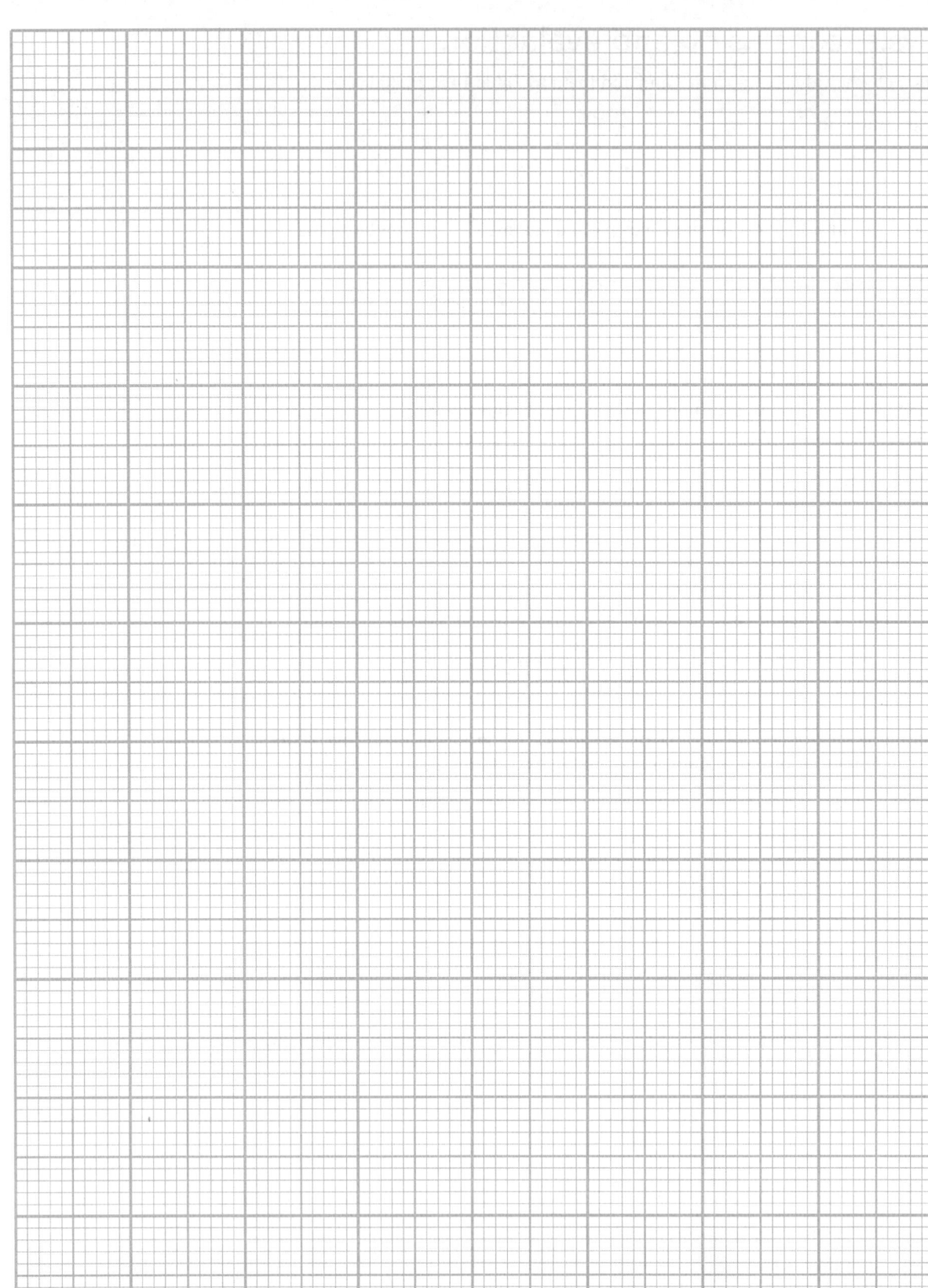

h   Explain how *L* may be determined accurately.

....

....

i   Explain how the measurements of *V* could be determined accurately.

....

....

....

....

Practical investigation 14.5: Observing charged particles

Your teacher will demonstrate this investigation.

When a beam of electrons enters a uniform magnetic field at right angles to the direction of the field, the electrons are deflected in a circular direction as shown in Figure 14.6. This experiment may be demonstrated using an electron tube. A pair of Helmholtz coils can be used to create a uniform magnetic field.

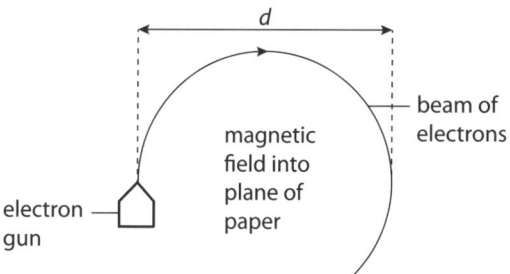

Figure 14.6: Path of electrons in a magnetic field.

Safety considerations

* Make sure you have read the Safety advice at the beginning of this book and listen to any advice from your teacher before carrying out this investigation.

* High voltages are being used so there should not be any bare wire connections.

Method

1   The apparatus is set up to produce a beam of electrons that is deflected, as shown in Figure 14.6.

2   Measure and record the voltage V in Table 14.3 in the Results section.

3   Measure the diameter d of the beam of electrons. Record your results in Table 14.3. Include the absolute uncertainty in d.

4   Change the voltage and repeat steps **2** and **3** until you have six different voltage values.

Results

	±	±
	±	±
	±	±
	±	±
	±	±
	±	±

Table 14.3: Results table.

Analysis, conclusion and evaluation

a   Calculate values of d^2 / cm^2 and add them to Table 14.3. Include the uncertainties in d^2.

b   Draw a graph of d^2/ cm^2 on the y-axis plotted against V / V on the x-axis.

c   It is suggested that the relationship between d^2 and V is:

$$\left(\frac{d}{2}\right)^2 = \frac{2mV}{B^2e}$$

where m is the rest mass of an electron, e is the charge on an electron and B is the magnetic flux density.

Using this relationship, determine the gradient of your graph of d^2 against V in terms of m, B and e.

Gradient = ...

d   Use the uncertainty in your values of d^2 to draw error bars on your graph. Draw a straight line of best fit and a worst acceptable straight line.

> **TIP**
>
> Remember the rule for combining uncertainties with power terms.

e Determine the gradient of the line of best fit and the gradient of the line of worst fit. You do not need to give units. Estimate the uncertainty in your value of the gradient.

Gradient of best-fit line = ...

Gradient of worst-fit line = ...

Uncertainty in gradient = ...

f Using your value of the gradient, determine a value for B. Include an appropriate unit for B.

B = ...

g Determine the percentage uncertainty for B.

Percentage uncertainty in B = ...%

h Describe a method to reduce the uncertainty in d.

....

....

....

....

Chapter 15

Electromagnetic induction and alternating currents

This chapter relates to Chapter 26: Electromagnetic induction and Chapter 27: Alternating currents, in the coursebook.

In this chapter you will complete investigations on:

- 15.1 Planning an investigation into the height of a metal ring above a current-carrying coil
- 15.2 A bar magnet moving through a coil
- 15.3 Planning an investigation into eddy currents
- 15.4 Planning an investigation into the effect of the iron core of a transformer
- 15.5 Ripple voltages in a rectification circuit.

Practical investigation 15.1: Planning
The height of a metal ring above a current-carrying coil

When current flows through a coil, as shown in Figure 15.1, the aluminium ring rises a height h.

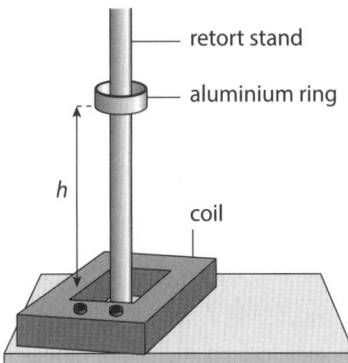

retort stand

aluminium ring

h

coil

Figure 15.1: Current flowing through a coil and aluminium ring rising to a height on a retort stand.

It is suggested that the relationship between the height h and the current I is given by:

$h = pI^q$

where p and q are constants.

You are going to design a laboratory experiment to test the relationship between h and I. In your account you will:

- write an account of the procedure to be followed

- describe the measurements to be taken

- describe the types of variables involved

- describe how the data can be analysed to determine values for p and q

- give one or two safety precautions that may be taken.

Variables

List the dependent variable, the independent variable and the variables that should be controlled. The variables to be controlled are quantities that must be kept the same.

- Dependent variable: ....

- Independent variable: ....

- Variables to be controlled: ....

   ....

YOU WILL NEED

List the equipment you will need and draw a labelled diagram of how you will set up the apparatus to obtain the necessary measurements.

- ....

- ....

- ....

- ....

- ....

- ....

- ....

TIP

What type of current is needed?

Safety considerations

- Make sure you have read the Safety advice at the beginning of this book and listen to any advice from your teacher before carrying out this investigation.

- ....

  ....

Method

Describe how you will carry out the experiment.

....

....

....

....

....

....

....

....

....

....

> **TIP**
>
> Explain clearly how *h* may be determined accurately.

Results

Draw a table of results which can be used to record and process the data from this experiment. You do not have to fill in the table. Remember to include the correct units in the column headings.

Analysis, conclusion and evaluation

a   Describe how you can analyse the data to show the relationship between h and I. Your account should include plotting a graph and using the gradient and y-intercept of the graph to determine values for p and q.

....

....

....

....

....

Practical investigation 15.2: A bar magnet moving through a coil

A magnet is placed on a trolley on an inclined plane. The trolley is released and travels towards a coil. As the trolley passes through the coil, the maximum induced electromotive force (e.m.f.) is measured. The speed of the trolley and magnet is determined by using a light gate connected to a timer immediately prior to the trolley entering the coil.

YOU WILL NEED

Equipment:

- bar magnet • coil (50–100 turns) • trolley • inclined plane • light gate
- voltmeter or cathode ray oscilloscope • timer • ruler • connecting leads
- retort stands.

Safety considerations

- Make sure you have read the Safety advice at the beginning of this book and listen to any advice from your teacher before carrying out this investigation.

- There are no specific safety issues for this investigation.

Method

1   Set up the apparatus as shown in Figure 15.2.

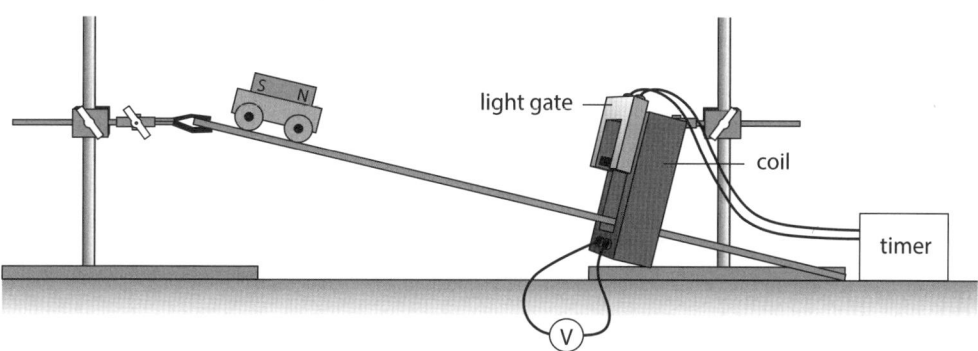

Figure 15.2: Trolley on inclined slope and light gate.

2   Measure and record the length L of the magnet in the Results section. Include the absolute uncertainty in L.

3   Release the trolley.

4   Measure and record the maximum reading E on the voltmeter. Measure and record the time t from the timer. Record your results in Table 15.1 in the Results section. Repeat the experiment for the same initial height of the trolley.

5   Change the initial height of the trolley and repeat steps **3** and **4** until you have six different values of induced e.m.f. E.

Results

$L =$ ....

t_1 / ms	t_2 / ms	average t /ms	$\frac{1}{t}$ / ms^{-1}	E / mV
		±	±	
		±	±	
		±	±	
		±	±	
		±	±	
		±	±	

Table 15.1: Results table.

> **TIP**
>
> Remember to include all your raw data in the table of results.

Analysis, conclusion and evaluation

a   Calculate the values of average t / ms and add them to Table 15.1. Include the uncertainties in average t.

b   Calculate the values of $\frac{1}{t}$ / s^{-1} and add them to Table 15.1. Include the uncertainties in $\frac{1}{t}$.

c   Draw a graph of E / mV on the y-axis plotted against $\frac{1}{t}$ / s^{-1} on the x-axis.

d   It is suggested that the relationship between E and $\frac{1}{t}$ is:

$$E = k\frac{L^2}{t}$$

where k is a constant.

Using the equation, determine the gradient of your graph of E against $\frac{1}{t}$ in terms of k.

Gradient = ...

e   Use the uncertainty in your values of $\frac{1}{t}$ to draw error bars on your graph. Draw a straight line of best fit and a worst acceptable straight line.

f   Determine the gradient of the line of best fit and the gradient of the line of worst fit. You do not need to give units. Estimate the uncertainty in your value of the gradient.

Gradient of best-fit line = ...

Gradient of worst-fit line = ...

Uncertainty in gradient = ...

> **TIP**
>
> Think about how to determine the uncertainties in t when you have two values.

g Using your value of the gradient, determine a value for k. Include appropriate units for k.

$k =$ ....

h Determine the percentage uncertainty for k.

Percentage uncertainty in $k =$ ....%

i Determine the absolute uncertainty in k.

Absolute uncertainty in $k =$ ....%

j Describe difficulties in the determination of t and explain how the experiment could be improved.

....

....

....

k Explain how the measurement of E could be improved.

....

....

Practical investigation 15.3: Planning Eddy currents

A thin sheet of copper is cut so that it has a small gap, as shown on the left in Figure 15.3.

 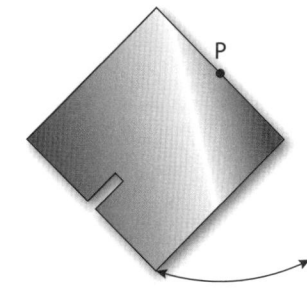

Figure 15.3: Copper sheet with slot in two positions.

The sheet is pivoted at P and placed in a magnetic field. It is displaced as shown on the right in Figure 15.3 and released. The time taken for the sheet to come to a rest is measured.

It is suggested that the relationship between the time t for the sheet to come to a rest and the distance d is given by:

$t = ke^{-qd}$

where k and q are constants.

You are going to design a laboratory experiment to test the relationship between t and d. In your account you will:

- write an account of the procedure to be followed
- describe the measurements to be taken
- describe the types of variables involved
- describe how the data is analysed to determine values for k and q
- give one or two safety precautions that may be taken.

Variables

List the dependent variable, the independent variable and the variables that should be controlled. The variables to be controlled are quantities that must be kept the same.

- Dependent variable: ....

- Independent variable: ....

- Variables to be controlled: ....

  ....

YOU WILL NEED

List the equipment you will need and draw a labelled diagram of how you will set up the apparatus to obtain the necessary measurements.

- ....
- ....
- ....
- ....
- ....

Safety considerations

- Make sure you have read the Safety advice at the beginning of this book and listen to any advice from your teacher before carrying out this investigation.

- ....
- ....

Method

Describe how you will carry out the experiment.

....

....

....

....

....

....

....

....

....

Results

Draw a table of results which can be used to record and process the data from this experiment. You do not have to fill in the table. Remember to include the correct units in the column headings.

Analysis, conclusion and evaluation

a   Describe how you can analyse the data to show the relationship between t and d. Your account should include plotting a graph and using the gradient and y-intercept of the graph to determine values for k and q.

....

....

....

....

Practical investigation 15.4: Planning The effect of the iron core of a transformer

A transformer usually has an iron core. Figure 15.4 shows an iron C-core with an iron bar on the top.

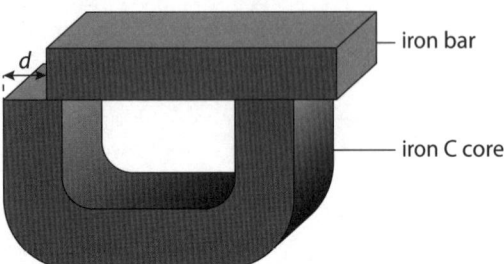

Figure 15.4: Iron C-core with iron bar on top.

A student is investigating how the electromotive force (e.m.f.) E across the secondary coil of a transformer varies with the distance d of the iron bar as shown.

It is suggested that:

$E = E_0 e^{-kd}$

where E_0 and k are constants.

You are going to design a laboratory experiment based on Figure 15.4 to test the relationship between E and d. In your account you will:

- write an account of the procedure to be followed
- describe the measurements to be taken
- describe the types of variables involved
- describe how the data is analysed
- give one or two safety precautions that may be taken.

Variables

List the dependent variable, the independent variable and the variables that should be controlled. The variables to be controlled are quantities that must be kept the same.

- Dependent variable: ....

- Independent variable: ....

- Variables to be controlled: ....

  ....

YOU WILL NEED

List the equipment you will need and draw a labelled diagram of how you will set up the apparatus in order to obtain the necessary measurements.

- ....
- ....
- ....
- ....
- ....
- ....
- ....
- ....

TIP

Think about how to measure a small distance d accurately.

Safety considerations

- Make sure you have read the Safety advice at the beginning of this book and listen to any advice from your teacher before carrying out this investigation.

- ....

- ....

Method

Describe how you will carry out the experiment.

....

....

....

....

....

....

....

....

....

....

Results

Draw a table of results which can be used to record and process the data from this experiment. You do not have to fill in the table. Remember to include the correct units in the column headings.

Analysis, conclusion and evaluation

a   Describe how you can analyse the data to test the relationship between E and d. Your account should include plotting a graph and using the gradient and y-intercept of the graph to determine values for E_0 and k.

....

....

....

Practical investigation 15.5:
Ripple voltages in a rectification circuit

An alternating current is **rectified** by using a diode. A capacitor is placed in parallel with the resistor to smooth the output. The ripple voltage V_R is defined as the difference between the maximum and minimum output voltage.

YOU WILL NEED

Equipment:

- capacitor • diode • various resistors in the range 3.3 kΩ to 10 kΩ
- a.c. power supply • connecting leads • cathode ray oscilloscope.

Safety considerations

- Make sure you have read the Safety advice at the beginning of this book and listen to any advice from your teacher before carrying out this investigation.

- There are no specific safety issues for this investigation.

Method

1 Set up the circuit shown in Figure 15.5.

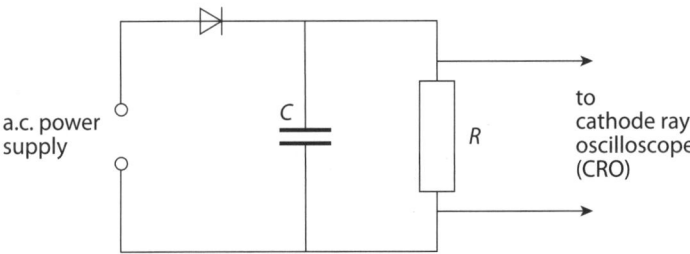

Figure 15.5: Circuit diagram.

2 Measure and record the minimum voltage V_{min} from the oscilloscope in Table 15.2 in the Results section. Include the absolute uncertainty in V_{min}.

3 Change the value of the resistor R and repeat step **2** until you have six different values of V_{min} and R.

4 Disconnect the resistor. Measure and record the peak voltage V_{max} from the oscilloscope in the Results section.

5 Measure and record the frequency f of the alternating power supply in the Results section.

Results

Frequency, f = ... V_{max} = ...

Table 15.2: Results table.

> **TIP**
>
> To determine frequency from an oscilloscope, the period T of the waveform needs to be determined.

Analysis, conclusion and evaluation

a Calculate the values of $\frac{1}{R}$ / $10^{-3}\,\Omega^{-1}$ and add them to Table 15.2. Include the uncertainties in $\frac{1}{R}$.

b Calculate the values of V_R / V and add them to Table 15.2. Include the uncertainties in V_R.

> **KEY EQUATION**
>
> $V_R = V_{max} - V_{min}$

> **TIP**
>
> Think about how to combine uncertainties when subtracting two quantities.

c Draw a graph of V_R / V on the y-axis plotted against $\frac{1}{R}$ / $10^{-3}\,\Omega^{-1}$ on the x-axis.

d It is suggested that the relationship between V_R and $\frac{1}{R}$ is:

$$V_R = \frac{V_{max}}{fCR}$$

where C is the capacitance of the capacitor.

Using this relationship, determine the gradient of your graph of V_R against $\frac{1}{R}$ in terms of C.

Gradient = ...

e Use the uncertainty in your values of V_R / V to draw error bars on your graph. Draw a straight line of best fit and a worst acceptable straight line.

f Determine the gradient of the line of best fit and the gradient of the line of worst fit. You do not need to give units. Estimate the uncertainty in your value of the gradient.

Gradient of best-fit line = ...

Gradient of worst-fit line = ...

Uncertainty in gradient = ...

g Using your value of the gradient, determine a value for C. Include appropriate units for C.

C = ...

h Determine the percentage uncertainty in C.

Percentage uncertainty in C = ...%

i Determine the absolute uncertainty in C.

Absolute uncertainty in C = ...%

j Explain how R could be accurately determined.

....

....

k Explain how the uncertainty in values of V_R could be reduced.

....

....

Chapter 16

Quantum physics, nuclear physics and medical imaging

CHAPTER OUTLINE

This chapter relates to Chapter 28: Quantum physics, Chapter 29: Nuclear physics and Chapter 30: Medical imaging, in the coursebook.

In this chapter you will complete investigations on:

- 16.1 Determining Planck's constant

- 16.2 Measuring a radioactive decay constant

- 16.3 Planning an investigation into X-ray attenuation.

Practical investigation 16.1: Determining Planck's constant

The current in a light-emitting diode (LED) is a series of moving electrons. Within the LED, the energy of an electron is transferred to the energy hf of a photon. If there is a potential difference V across an LED then the maximum energy of the electron is eV.

It is suggested that, if this energy is just large enough to cause the emission of the photon, then:

$$eV = hf = \frac{hc}{\lambda}$$

where c is the velocity of light, e is the charge on an electron, h is **Planck's constant**, f is the frequency of the light and λ is the wavelength of light.

You will use several LEDs of different colours to investigate the relationship between the voltage at which the LEDs start to emit light and the wavelength of the light that they emit. You will also use your results to find a value for Planck's constant h.

> **KEY WORDS**
>
> **Planck's constant:** a fundamental constant which links the energy of a photon E and its frequency f

> **KEY EQUATION**
>
> **Planck's constant,**
> $E = hf$

> **YOU WILL NEED**
>
> **Equipment:**
> • low voltage d.c. power supply • several LEDs of different colours • safety resistor of a few hundred ohms • variable resistor used as a potentiometer • digital multimeter or voltmeter • small opaque tube, e.g. black card, to fit over an LED.
>
> **Access to:**
> • the internet or a colour chart showing the wavelength of light of different colours.

Safety considerations

- Make sure you have read the Safety advice at the beginning of this book and listen to any advice from your teacher before carrying out this investigation.

- There are no special safety issues in this experiment since the voltages and currents are small.

- It is sensible to make sure there is a safety resistor connected in series with each LED. Without this resistor it is easy to blow the LED, which is not likely to be dangerous but will *destroy* the LED.

Method

1   Set up the circuit as shown in Figure 16.1. Start with a low potential difference across the LED and slowly increase the potential difference until the LED just begins to glow.

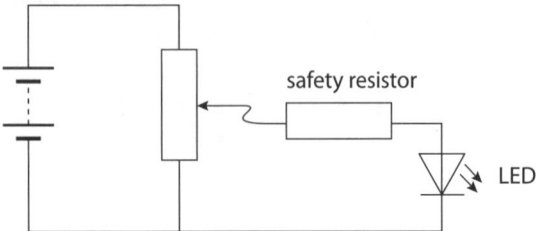

Figure 16.1: Circuit diagram.

This is best done in a darkened part of the room, placing a black tube over the LED and looking down the tube at the LED. If you are working in a group, each person should attempt each measurement, as it is difficult to estimate when the LED just starts to emit light.

2   Note the voltmeter reading V_{min} when you just start to see the LED emit light. Record your reading in Table 16.1 in the row for LED 1. Repeat your reading and obtain the average value and the uncertainty in V_{min}. Record the values in Table 16.1 in the Results section.

> **TIP**
>
> The uncertainty in V is half the difference between the largest and smallest values of V. If all readings are the same then it is the smallest scale division on the voltmeter.

3 Replace the LED with one of a different colour and repeat your measurement of the minimum voltage V_{min}. Obtain the average value and uncertainty. Record all your readings in Table 16.1. Space has been left for five different LEDs but you may not have this number available.

4 Estimate the wavelength λ of the light emitted by each LED. This can be done by comparing the colour of the light emitted by your LED with a chart obtained from the internet.

Write your estimated values of the wavelength in Table 16.1 and, if you can, estimate each uncertainty.

Results

LED	Voltmeter reading V_{min} / V				λ / m	$\frac{1}{\lambda}$ / m⁻¹
	1st	2nd	3rd	average		
1					___ × 10⁻⁷	
				±	±	
2					___ × 10⁻⁷	
				±	±	
3					___ × 10⁻⁷	
				±	±	
4					___ × 10⁻⁷	
				±	±	
5					___ × 10⁻⁷	
				±	±	

Table 16.1: Results table.

> **TIP**
>
> Write the uncertainty after the ± symbol. Do not worry if you cannot do this for the wavelength.

Analysis, conclusion and evaluation

a Complete Table 16.1 by calculating values of $\frac{1}{\lambda}$ / m⁻¹ and, if possible, their uncertainties.

b Plot a graph of your average value of V_{min} / V on the y-axis against λ^{-1} / m⁻¹ on the x-axis. Include error bars along both axes, using the uncertainties in Table 16.1.

Draw the straight line of best fit and a worst acceptable straight line on your graph.

c   Calculate the gradient of your best-fit line and your worst-fit line. Give each
    gradient a unit, which you should calculate from the units in Table 16.1.

    Gradient of best-fit line = ...

    Gradient of worst-fit line = ...

d   Using the expression $eV = \dfrac{hc}{\lambda}$, state an expression for the gradient of your graph in
    terms of e, h and c.

    Gradient = ...

e   Use your value for the gradient of the best-fit line to calculate a value for Planck's
    constant h, given that $e = 1.6 \times 10^{-19}$ C and $c = 3.0 \times 10^8$ m s^{-1}. Give a unit for your
    answer.

    h = ...

f   Use your value for the gradient of the worst-fit line to calculate a value for the
    uncertainty in h.

    Uncertainty = ...

g   Compare the value you have obtained in part **e** with the accepted value for
    Planck's constant.

    ....

    ....

h   If the accepted value does not lie within your limits of uncertainty then there may
    be a systematic error. Suggest some causes of this error or some causes of the
    uncertainty in your values of the voltmeter reading V.

    ....

    ....

    ....

    ....

    ....

    ....

Practical investigation 16.2: Data analysis
Measuring a radioactive decay constant

KEY WORDS

radioactive decay constant: the probability that an individual nucleus will decay per unit time interval

In this investigation you will use the results of an experiment on radioactive decay to measure the **radioactive decay constant** of an isotope of protactinium.

Figure 16.2 shows the experimental arrangement.

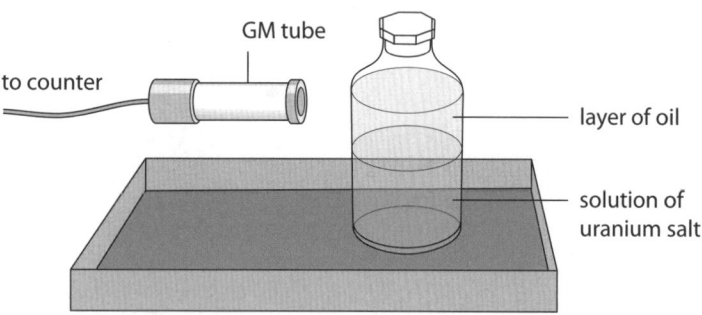

Figure 16.2: Radiation counter and bottle containing solution of uranium salt.

The plastic bottle contains a salt of uranium dissolved in water. One of the decay products of the uranium is protactinium, which is the only decay product that dissolves in oil. When the bottle is shaken, the protactinium in the solution dissolves in the oil and rises to the surface. Protactinium is also radioactive. The beta particles that it emits pass through the plastic bottle.

After the oil has settled at the top, the number of counts recorded in a time of 10 s is measured and this is repeated every 20 s. The readings are shown in Table 16.2 in the Results section. After a long time, when all the protactinium in the layer has decayed, the background count is taken for a time of 100 s. This background reading is also shown in the Results section.

It is not possible to repeat the experiment starting with the same initial count and so the uncertainty has been calculated as the square root of the reading, which is a suitable estimate in this case.

Safety considerations

* Make sure you have read the Safety advice at the beginning of this book and listen to any advice from your teacher before carrying out this investigation.

* What precautions would you take to make sure that the experiment is performed safely?

....

....

....

Results

Background count taken over 100 s = 50

Time t / s	Count in 10 s	Corrected count rate C / s⁻¹	ln (C / s⁻¹)
0	123 ± 11	11.8 ± 1.1	2.47 ± 0.09
20	100 ± 10	± 1.0	±
40	80 ± 9	± 0.9	±
60	68 ± 8	± 0.8	±
80	52 ± 7	± 0.7	±
100	45 ± 7	± 0.7	±
120	34 ± 6	± 0.6	±
140	31 ± 6	± 0.6	±

Table 16.2: Results table.

> **TIP**
>
> Take care: these logarithms are to base e; *not* to base 10.

Analysis, conclusion and evaluation

a   Calculate the background count for a time of 10 s and use your value to find the corrected count rate C in one second for all the readings in Table 16.2. One value has been inserted for you.

b   Calculate values for ln (C / s⁻¹) and the uncertainty. Record the values in Table 16.2.

> **TIP**
>
> Check that the first row calculations for the corrected count rate and ln (C / s⁻¹) are correct. This will show that you know how to find both values correctly.

> **TIP**
>
> The uncertainty in ln C is easily found by finding ln C of the largest count, e.g. 11.8 + 1.1 and subtracting.

c   Plot a graph of $\ln(C/\mathrm{s}^{-1})$ on the y-axis against time t/s on the x-axis, using the grid provided. Insert error bars for $\ln(C/\mathrm{s}^{-1})$. Draw straight lines of best fit and worst fit on your graph.

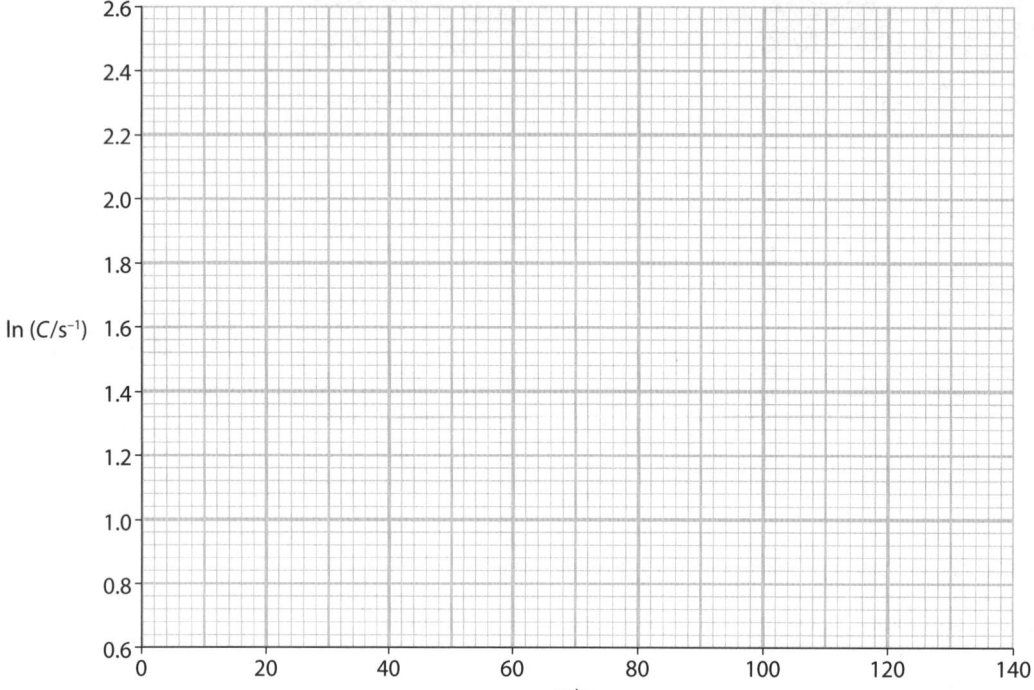

d   Determine the gradients of the line of best fit and the line of worst fit.

Gradient of line of best fit = ……………………..

Gradient of line of worst fit = ……………………

e   Theory suggests that the corrected count rate C decreases exponentially with time t according to the expression:

$C = C_0 e^{-\lambda t}$

where λ is the radioactive decay constant.

By taking logarithms to base e, complete the expression:

$\ln C =$ ……………………

f Use your value for the gradients to determine a value for λ and its uncertainty. Give the unit of your answer.

Radioactive decay constant λ = ... $\pm$ ....

Unit of answer = ...

g A scientist repeats the experiment without using the oil to dissolve the protactinium in the uranium. Describe why the experiment does *not* work.

....

....

....

....

Practical investigation 16.3: Planning X-ray attenuation

A medical physicist wishes to show that the intensity of X-rays decreases exponentially as the thickness of a lead absorber increases and to measure the **attenuation coefficient** for lead.

The apparatus that they use is shown in Figure 16.3, and contains an X-ray source and a Geiger–Muller (GM) tube connected to a counter. A piece of lead is placed between the GM tube and the X-ray source. The counter gives a measure of the number of X-ray photons entering the tube in the time for which the count is made. The physicist knows that about 5 mm of lead halves the count rate for these X-rays.

> **KEY WORDS**
>
> **attenuation coefficient:** the fraction of a beam of X-rays that is absorbed per unit thickness of the absorber

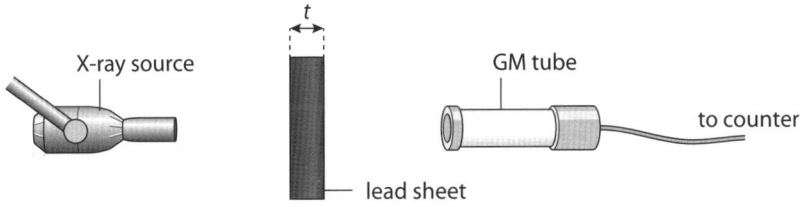

Figure 16.3: X-ray source, lead sheet and radiation counter.

It is suggested that the count rate C measured by the GM tube is related to the thickness t of the lead sheet by the equation:

$C = C_0 e^{-\alpha t}$

where α is a constant, the attenuation coefficient, and C_0 is the count rate with no absorber.

You are going to write an account of the procedure the physicist should follow, being careful to:

- state what readings are taken
- describe how to allow for any background radiation that enters the GM tube
- name the instrument used to measure t
- give details of the thicknesses of lead to be used
- describe how the count rate is measured accurately
- give a safety precaution the physicist should take.

> ### KEY EQUATIONS
>
> attenuation of ultrasound and X-rays in matter $C = C_0 e^{-\alpha t}$

> ### TIP
>
> Count rate is the count in a certain time interval, often a second, minute or hour.

Variables

List the dependent variable, the independent variable and the variables that should be controlled. The variables to be controlled are quantities that must be kept the same.

- Dependent variable: ....
- Independent variable: ....
- Variables to be controlled: ....

  ....

> ### YOU WILL NEED
>
> List the equipment needed, apart from that shown in Figure 16.3. Include the name of the instrument to be used to measure t.
>
> - ....
> - ....

Safety considerations

Give a safety precaution the physicist should take.

- ....

Method

Describe how the physicist should carry out the experiment; in particular:

1   State what readings are taken.

2   Describe how to allow for any background radiation that enters the GM tube.

3   Give details of the thicknesses of lead to be used.

4   Draw columns for a results table with headings to show what quantities are measured and calculated.

5   Describe how the count rate is measured accurately.

6   Give some extra detail to ensure that the measurements are accurate.

....

....

....

....

....

....

....

....

....

....

....

....

....

....

Analysis, conclusion and evaluation

a   Given that $C = C_0 e^{-\alpha t}$, complete the equation for $\ln C$.

$\ln C = $ ....

b   Decide which graph produces a straight line if the expression for C is correct. Write down which quantity is plotted on the x-axis and which quantity on the y-axis.

x-axis .... y-axis ....

c   Draw a rough sketch of the graph you expect on the axes.

> **TIP**
>
> Label the axes of your graph and include their units.

d   Using terms in the equation in part **a**, determine the gradient and intercept of your graph.

Gradient = ...   Intercept = ...

e   Another material is used instead of lead, and the experiment is repeated. The new material has an attenuation coefficient α that is half that of lead. Describe how the graph obtained differs from the original graph.

....

....

....

....

> Chapter 17
Astronomy and cosmology

Practical investigation 17.1:
Data analysis
Stefan's law

A **black body** absorbs electromagnetic radiation of all wavelengths that fall on it. It is also a more efficient emitter of radiation than any other body. Radiation emitted from a black body is called black-body radiation.

Stefan's law states that, for a perfect black body of surface area A at thermodynamic temperature T (measured in K), the luminosity L or total power of radiation emitted is given by:

$L = \sigma A T^n$

where $n = 4.0$ and σ is the Stefan–Boltzmann constant.

A small hole in an electrically heated furnace is a good emitter of radiation. The temperature of the furnace can be measured separately, often with a thermocouple thermometer.

The total power P radiated from a small hole in an electrically heated furnace is given by:

$P = \varepsilon \sigma A T^n$

where ε is a constant known as the emissivity of the furnace.

The emissivity measures how closely the radiation emitted from the hole approaches that from a perfect black body.

A scientist takes measurements from an experiment with such a furnace to test the relationship between the power emitted and the temperature of the furnace and to measure the emissivity of the furnace.

Figure 17.1 shows the apparatus used. The radiometer measures the net amount of thermal power received from the small hole in the furnace. If the temperature of the furnace is much higher than the temperature of the radiometer, the amount of thermal

KEY WORDS

black body: an idealised object that absorbs all incident electromagnetic radiation falling on it. It has a characteristic emission spectrum and intensity that depends only on its thermodynamic temperature

KEY EQUATION

Stefan's law
$L = 4\pi\sigma r^2 T^4$

power emitted by the radiometer can be ignored. The radiometer output is calibrated to measure directly the power in watts emitted by the small hole in the furnace at the particular distance from the furnace.

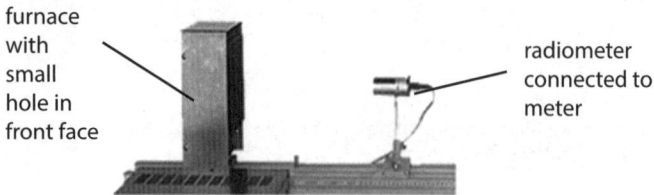

furnace with small hole in front face

radiometer connected to meter

Figure 17.1: Furnace with a small hole in front face directed towards a radiometer connected to meter.

Values for the measured power P at different temperatures T of the furnace are shown in Table 17.1 in the Results section. The values for P show the uncertainty obtained from several readings.

Results

Temperature T of furnace / °C	Power from hole P / W	T / K	lg $(T$ / K$)$	lg $(P$ / W$)$
1400	0.9 ± 0.2	1670	3.223	-0.05 ± 0.10
1600	1.5 ± 0.2			
1800	2.2 ± 0.2			
2000	3.0 ± 0.2			
2200	4.1 ± 0.2			
2400	5.7 ± 0.2			

Table 17.1: Results table.

Analysis, conclusion and evaluation

a   List the dependent variable, the independent variable and one variable to be controlled. The variable to be controlled is a quantity that must be kept the same.

   •   Dependent variable ....

   •   Independent variable ....

   •   Variable to be controlled ....

b   By taking logs to base 10 of the formula $P = \varepsilon \sigma A T^{n}$, obtain a formula for lg P in terms of lg T and ε, σ, A and n.

   lg P = ... × lg T + ...

c Complete Table 17.1 by calculating values for T / K, $\lg(T / K)$ and $\lg(P / W)$. Include values for the uncertainty in the values for $\lg(P / W)$. One row has been done for you.

d Plot a graph of $\lg(P / W)$ on the y-axis against $\lg(T / K)$ on the x-axis using the graph grid. Include error bars for $\lg(P / W)$. Draw a straight line of best fit and a straight line of worst fit on your graph.

> **TIP**
>
> Remember,
> $T / K = T / °C + 273$
>
> lg means log to base 10.

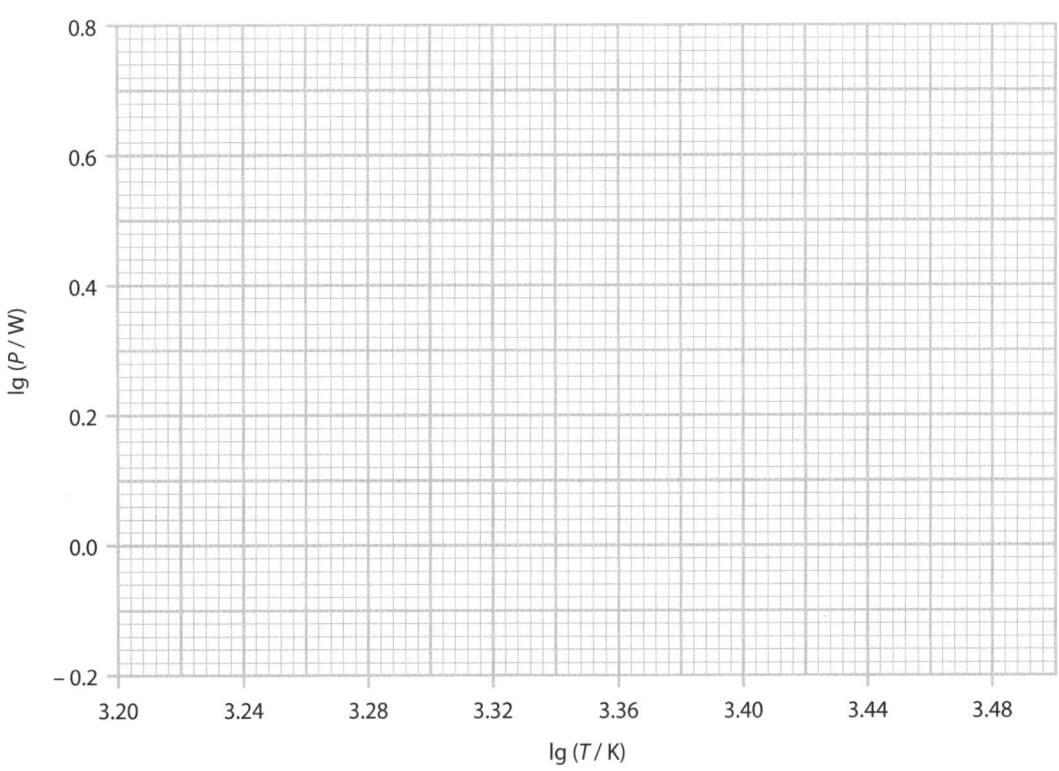

e Determine the gradients of the line of best fit and the line of worst fit.

Gradient of line of best fit = ……………..

Gradient of line of worst fit = …………….

f Use your values for the gradients to determine a value for n and its uncertainty.

$n =$ ………………

g Choose a value of $\lg T$ in the range of values on your graph and read off the corresponding value of $\lg P$ from your line of best fit.

Use these values in your equation in **b** to obtain a value for ε, the emissivity of the furnace surface.

Take the Stefan–Boltzmann constant σ as 5.67×10^{-8} W m^{-2} K^{-4} and the area A of the hole in the furnace from which the radiation is emitted as 2.0×10^{-6} m^2.

$\varepsilon =$ …………….

h   Compare the value you have obtained for *n* with the value in Stefan's law. Do the two values agree within the limits of experimental uncertainty?

....

i   Use your answer to **g** to suggest why the radiation from the hole in the furnace can be considered to be black body radiation.

....

....

j   Suggest why the percentage uncertainty in the power is largest at lower temperatures.

....

....

....

....

Practical investigation 17.2: Planning Wien's displacement law

Figure 17.2 shows a graph of the spectrum of the radiation emitted by a black body at four different temperatures.

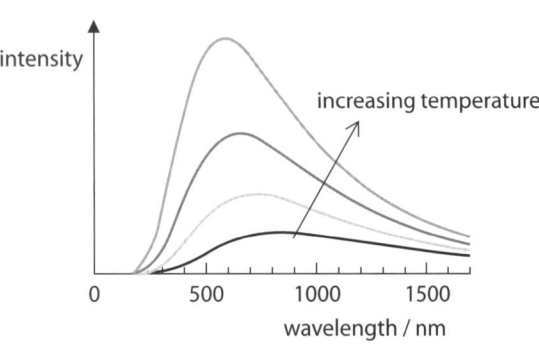

Figure 17.2: Graph showing sketches of the spectrum of the radiation emitted by a black body at four different temperatures, with direction of increasing temperature shown by an arrow.

It is suggested that the wavelength of the peak of the emission λ_{max} is inversely proportional to the thermodynamic temperature T and that the formula relating these quantities is:

$$\lambda_{max} \times T = \text{constant}$$

You are going to design a laboratory experiment to test the relationship between λ_{max} and T for radiation from a furnace.

You have available a furnace with a small hole and a radiometer, as shown in Figure 17.3. The furnace can operate over temperatures from 1000 to 2500 °C. The radiometer measures the total power of all relevant wavelengths that enter a small slit placed in front of it.

You also have available a diffraction grating with a known grating spacing d and other, more common, laboratory equipment.

You may have to remind yourself how to use a diffraction grating to measure wavelength (see Chapter 13 of the coursebook) and think about how to produce a narrow beam of radiation from the furnace.

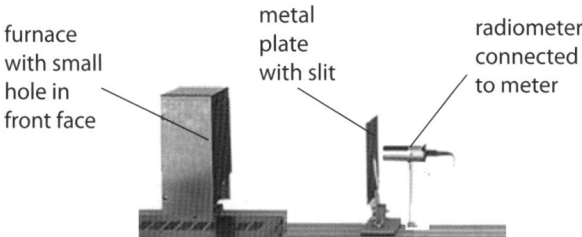

furnace with small hole in front face

metal plate with slit

radiometer connected to meter

Figure 17.3: As Figure 17.1, with a metal plate with slit placed between the furnace and the radiometer.

You are asked to write an account of the procedure, including:

- the procedure to be followed
- the measurements to be taken
- the variables involved
- how the data is to be analysed
- one or two safety precautions that may be taken.

Variables

List the dependent variable, the independent variable and the variables that should be controlled. The variables to be controlled are quantities that must be kept the same.

- Dependent variable: ....

- Independent variable: ....

- Variables to be controlled: ....

  ....

> **TIP**
>
> Radiation from the hole in the furnace spreads out in all directions. You will need to make the radiation into a beam to use the diffraction grating. This can be done with a slit – do not use a lens as glass will absorb some wavelengths. You will not be able to see the infra red from the source and so the radiometer will need to detect it.

YOU WILL NEED

List the equipment needed, apart from that shown in Figure 17.3. Be careful that any thermometer you choose is suitable for the temperature of the furnace. Draw a diagram showing how to set up the diffraction grating to obtain the necessary measurements.

- ....
- ....
- ....
- ....
- ....
- ....
- ....
- ....
- ....
- ....

Safety considerations

....

....

Method

Describe how to carry out the experiment, in particular:

- state what readings are taken and how the peak of the emission is found
- draw columns for a results table with headings to show what quantities are measured and calculated, with their units
- give some extra detail to ensure that the measurements are accurate
- describe how the wavelength of the peak emission can be found from your readings.

....

....

....

....

....

....

....

....

....

....

Analysis, conclusion and evaluation

a Given that $\lambda_{max} \times T = $ constant, decide which graph involving λ_{max} and T produces a straight line if the expression is correct.

Write down which quantity is plotted on the x-axis and which quantity on the y-axis.

x-axis: .... y-axis: ....

b Draw a rough sketch of the graph you expect on the axes.

> **TIP**
>
> Label the axes of your graph and include units.

c State how the constant in the expression $\lambda_{max} \times T = $ constant can be found from your graph.

....

d Suggest in which part of the electromagnetic spectrum the peak wavelength is found given that $\lambda_{max} \times T = 2.9 \times 10^{-3}$ m / K, using a sensible value of temperature to be found in a furnace.

....

....

Practical investigation 17.3:
Data analysis
Hubble's law

Hubble's law describes how the velocity v of stars and galaxies moving away from us increases with their distance d from Earth.

This law is expressed in the simple equation $v = Hd$, where H is what is known as the Hubble constant.

The velocity of stars and galaxies is determined from their **redshift**. The distances from the galaxies to Earth are determined using the standard candle properties of Cepheid variables and type 1a supernovae.

Table 17.2 shows the results for a number of different galaxies. The table contains estimates of the uncertainty in the distances.

Distances are shown in mega parsecs (Mpc), where 1 Mpc = 3.1×10^{19} km.

> **KEY WORDS**
>
> **Hubble's law:** the recession speed of a star or galaxy is directly proportional to its distance from Earth
>
> **redshift:** the increase in the wavelength of electromagnetic waves due to recession of the source

> **KEY EQUATION**
>
> **Hubble's law** $v = Hd$

Results

Galaxy	d / Mpc	v / km s⁻¹
NGC 0055	2.2 ± 0.2	129
NGC 0247	3.3 ± 0.6	195
NGC 0045	10 ± 3	467
NGC 0578	18 ± 4	1630
NGC 0063	18.8 ± 0.2	
NGC 0514	29 ± 4	2470
NGC 2271	32 ± 2	2600
NGC 0632	38 ± 4	3190

Table 17.2: Results table.

Analysis, conclusion and evaluation

a An astronomer measures the wavelength of a line in the hydrogen spectrum of the galaxy NGC 0063 as 659.13 nm. The wavelength of the same line from a laboratory-based source is 656.28 nm.

The formula for **redshift** is:

$$\frac{\Delta\lambda}{\lambda} \approx \frac{\Delta f}{f} \approx \frac{v}{c}$$

where $\Delta\lambda$ and Δf are the observed shifts in wavelength λ and frequency f as the source of an electromagnetic wave moves towards an observer at speed v.

Calculate the velocity of galaxy NGC 0063 and add your value to Table 17.2.

b Plot a graph of v / km s^{-1} on the y-axis against d / Mpc on the x-axis using the graph grid. Include error bars for d.

Draw straight lines of best fit and worst fit on your graph.

c Determine the gradient of the line of best fit and the gradient of the line of worst fit. Include the unit of your gradient.

Gradient of line of best fit = ……………..

Gradient of line of worst fit = …………….

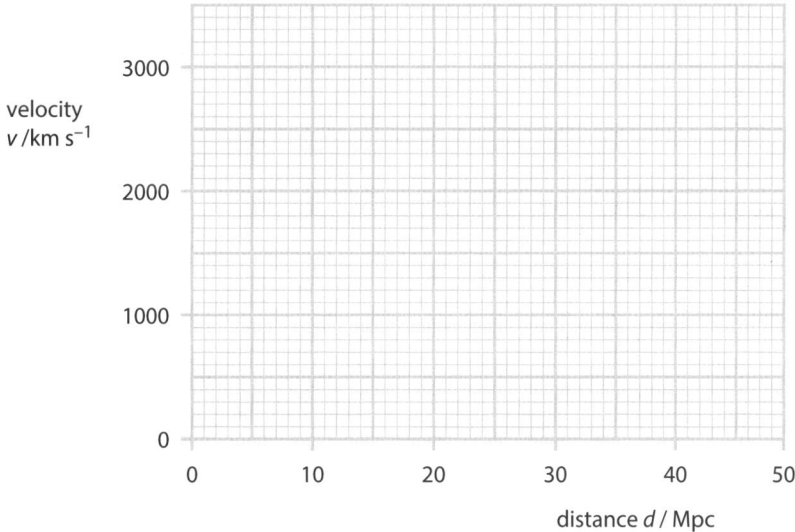

d Given that 1 Mpc = 3.1×10^{19} km, calculate the gradients in **c** in units of km s^{-1} ÷ km = s^{-1}.

Gradient of line of best fit = …………….. s^{-1}

Gradient of line of worst fit = …………….. s^{-1}

KEY EQUATION

redshift $\dfrac{\Delta\lambda}{\lambda} \approx \dfrac{\Delta f}{f} \approx \dfrac{v}{c}$

TIP

Your line should go through the origin. Do your best!

TIP

Don't forget to give units. You can use the units on the axes directly. You do not need to convert speeds to m s^{-1} or distances to m.

e According to Hubble's law, $v = Hd$, where H is the Hubble constant. Using your answer to **d**, obtain a value for the Hubble constant. Give your answer in s^{-1}.

Hubble constant = ... ± .... s^{-1}

f If one assumes that each galaxy has moved at a constant speed since the start of the Big Bang, then each galaxy has been moving for a time $T = \dfrac{d}{v} = \dfrac{1}{H}$

This time is known as the Hubble time and is an estimate of the age of the Universe.

Using your answer in **e**, obtain an estimate for the Hubble time. Give your answer in seconds and in billions of years.

T = ... seconds

= .... billions of years

> **TIP**
>
> 1 billion years = 10^9 years.

g Suggest any reasons why your line of best fit does not pass through all the error bars.

....

....

....

....

h If the Hubble constant is smaller than the value you have calculated in **f**, how does this affect the graph you have drawn in **b**? What does this tell you about the age of the universe?

....

....

....

....

....

⟩ Glossary

acceleration: the rate of change of velocity of an object

accuracy: an accurate measured value of a quantity is close to the true value of the quantity

amplitude: the maximum displacement of a particle or object from its equilibrium position

attenuation coefficient: the fraction of a beam of X-rays that is absorbed per unit thickness of the absorber

black body: an idealised object that absorbs all incident electromagnetic radiation falling on it. It has a characteristic emission spectrum and intensity that depends only on its thermodynamic temperature

Boyle's law: the pressure exerted by a fixed mass of gas is inversely proportional to its volume, provided the temperature of the gas remains constant

capacitance (of a capacitor): the charge stored on one plate per unit potential difference between the plates

centripetal acceleration: the acceleration of an object towards the centre of its circular motion

control variable: a quantity that has to be kept constant otherwise the relationship between the other variables is not tested fairly

damped oscillation: an oscillation in which frictional forces cause the energy of the system to be transferred to the surroundings

dependent variable: the variable in an experiment with a value that changes as the independent value is altered by the experimenter

electromotive force (e.m.f.): the amount of energy changed from other forms into electrical energy per unit charge produced by an electrical supply

equilibrium: when the resultant force and the resultant moment on a body are both zero

fiducial mark: a mark or marker used as a point of reference

frequency: the number of oscillations per unit time

gravitational potential: the work done per unit mass in bringing a mass from infinity to a point

gravitational potential energy: the energy a body has due to its position in a gravitational field

Hall voltage: the potential difference produced across the sides of a conductor when an external magnetic field is applied perpendicular to the direction of the current; the Hall voltage V_H is directly proportional to the magnetic flux density B

Hooke's law: provided the elastic limit is not exceeded, the extension of an object is proportional to the applied force

Hubble's law: the recession speed of a star or galaxy is directly proportional to its distance from Earth

independent variable: the variable in an experiment with a value that is altered by the experimenter

Kirchhoff's second law: the sum of the e.m.f.s around a closed loop is equal to the sum of the p.d.s in that same loop; this law represents the conservation of energy

line of best fit: straight line drawn as closely as possible to the points of a graph so that similar numbers of points lie above and below the line

magnetic field: a force field in which a magnet, a wire carrying a current, or a moving charge experiences a force

magnetic flux density: the force acting per unit current per unit length on a wire placed at right angles to the magnetic field

moment: the moment of a force about a point is the product of the force and perpendicular distance from the line of action of the force to the point

Newton's law of gravitation: any two point masses attract each other with a force that is directly proportional to the product of their mass and inversely proportional to the square of their separation

Newton's third law: when two bodies interact, the forces they exert on each other are equal and opposite

node: a point on a stationary wave where the amplitude is zero

ohm: the unit of electric resistance

percentage uncertainty: the absolute uncertainty as a fraction of the measured value

Planck's constant: a fundamental constant which links the energy of a photon E and its frequency f

potential difference (p.d.): the potential difference, V, between two points, A and B, is the energy transferred per unit charge as it moves from point A to point B

potential divider: a circuit in which two or more components are connected in series to a supply; the output voltage is taken across one of the components

power: the rate at which energy is transferred

precision: the smallest change in value that can be measured by an instrument or an operator. A precise measurement is one made several times, giving the same, or very similar, values; there is very little spread about the mean value

radioactive decay constant: the probability that an individual nucleus will decay per unit time interval

range: the difference between the largest value and the smallest value of a measurement

rectification: the process of converting alternating current (a.c.) into direct current (d.c.)

redshift: the increase in the wavelength of electromagnetic waves due to recession of the source

refraction: the change in direction of a wave as it crosses an interface between two materials where its speed changes

resistance: ratio of the potential difference across the component to the current in the component

resistivity: a measure of electrical resistance, defined as resistance × cross-sectional area / length

simple harmonic motion: motion of an oscillator in which its acceleration is directly proportional to its displacement from its equilibrium position and is directed towards that position

specific latent heat of vaporisation: the amount of heat energy per unit mass needed to convert unit mass of solid to liquid without change in temperature

stationary wave: a wave pattern produced when two progressive waves of the same frequency travelling in opposite directions combine

tensile: associated with tension or pulling, e.g. a tensile force

thermal energy: energy transferred from one object to another because of a temperature difference; another term for heat energy

thermistor: a device whose electrical resistance changes when its temperature changes

thermocouple: a device consisting of wires of two different metals across which an e.m.f. is produced when the two junctions of the wires are at different temperatures

uncertainty (also absolute uncertainty): an estimate of the spread of values around a measured quantity within which the true value will be found

upthrust: the force upwards in a liquid or gas caused by the pressure in the gas or liquid

velocity: an object's speed in a particular direction, or the rate of change of an object's displacement; it is a vector quantity

wavelength: the distance between two adjacent peaks or troughs in a wave or the distance between adjacent points having the same phase

Young modulus: the ratio of stress to strain for a given material, provided Hooke's law is obeyed